Survival of the Nicest

"This wonderful book could be read as a scientific explanation for a moral imperative to be kind to others. But it is so much more! Stefan Klein, an enticing storyteller, marshals the evidence for the value of altruism—not only to one's family but, much more interestingly, to one's self and one's tribe. Altruism is truly contagious!"

—ROALD HOFFMANN, Nobel Laureate, poet, and Frank H. T. Rhodes
Professor of Humane Letters Emeritus, Cornell University

"A scholarly tour de force about why generosity makes good sense, *Survival of the Nicest* is also compulsively readable. Klein argues convincingly that helping others is one of the best things we can do for ourselves."

—ELIZABETH SVOBODA, author of
What Makes a Hero?: The Surprising Science of Selflessness

"A thought-provoking and comprehensive review of the research on altruism, *Survival of the Nicest* validates humanistic principles and has far-reaching implications for today's world—especially for US politics and culture. An inspiration!"

—REBECCA HALE, president,
American Humanist Association, and co-owner of EvolveFISH.com

"An important contribution to the field of altruism and altruistic behavior and to a better and nicer world. I highly recommend this book."

—SAMUEL P. OLINER, PhD, Emeritus Professor of Sociology,
Humboldt State University, and founder and director, The Altruistic
Personality and Prosocial Behavior Institute

"In *Survival of the Nicest*, Stefan Klein poses three questions central to social science and ethics: (1) How is unselfishness possible? (2) What moves us to help others? And (3) why are some people more helpful than others? His wide-ranging answers to these questions suggest that altruism is born into us and that selflessness actually both makes us happy and will transform the world."

—KRISTEN RENWICK MONROE, Chancellor's Professor,
University of California, Irvine, and author of *The Heart of Altruism*

Survival of the Nicest

How Altruism Made Us Human and
Why It Pays to Get Along

STEFAN KLEIN

TRANSLATED BY DAVID DOLLENMAYER

THE EXPERIMENT

NEW YORK

For Elias

SURVIVAL OF THE NICEST: *How Altruism Made Us Human and Why It Pays to Get Along*
Copyright © Stefan Klein, 2010, 2014
Translation © David Dollenmayer, 2014
"The Mask of Evil," originally published in German as "Die Maske des Bösen."
Copyright 1943 by Bertolt-Brecht-Erben / Suhrkamp Verlag, from *Collected Poems of Bertolt Brecht*. Used by permission of Liveright Publishing Corporation. Translation for this edition by David Dollenmayer.

Illustrations on pages 31, 41, 126, and 144 courtesy of Karen Giangreco, after Hermann Hülsenberg, Berlin

Originally published in 2010 in German, in slightly different form, by S. Fischer Verlag GmbH, Frankfurt am Main.

The Experiment, LLC
220 East 23rd Street, Suite 301
New York, NY 10010–4674
www.theexperimentpublishing.com

The Experiment's books are available at special discounts when purchased in bulk for premiums and sales promotions as well as for fundraising or educational use. For details, contact us at info@theexperimentpublishing.com.

Many of the designations used by manufacturers and sellers to distinguish their products are claimed as trademarks. Where those designations appear in this book and The Experiment was aware of a trademark claim, the designations have been capitalized.

The Library of Congress has cataloged the hardcover edition as follows:

Klein, Stefan, 1965-
[Sinn des Gebens. English]
Survival of the nicest : how altruism made us human and why it pays to get along / Stefan Klein ; translated by David Dollenmayer.
    pages cm
Translation of: Der Sinn des Gebens : warum Selbstlosigkeit in der Evolution siegt und wir mit Egoismus nicht weiterkommen.
Includes bibliographical references and index.
ISBN 978-1-61519-090-4 (cloth) -- ISBN 978-1-61519-181-9 (ebook)
1. Altruism. 2. Cooperation. 3. Evolutionary psychology. I. Title.
BF637.H4K57 2014
155.7--dc23
2013024044

ISBN 978-1-61519-220-5
Ebook ISBN 978-1-61519-181-9

Cover design by Alison Forner
Cover image and image on pages 1 and 93 © Rossella Apostoli | Alamy and Sabri Deniz Kizil | Shutterstock
Text design by Pauline Neuwirth, Neuwirth & Associates, Inc.

Manufactured in the United States of America
Distributed by Workman Publishing Company, Inc.
Distributed simultaneously in Canada by Thomas Allen and Son Ltd.

First paperback printing September 2014
10 9 8 7 6 5 4 3 2 1

Contents

Preface   vii

Introduction   xi

PART I: YOU AND I

CHAPTER 1: The Unexplained Friendliness of the World   3

CHAPTER 2: Give and Take   20

CHAPTER 3: Building Trust   35

CHAPTER 4: Feelings Without Borders   54

CHAPTER 5: There Is Only One Love   73

PART II: ALL OF US

CHAPTER 6: Humans Share, Animals Don't   95

CHAPTER 7: It's the Principle of the Thing   116

CHAPTER 8: Us Against Them   137

CHAPTER 9: The Evil in Goodness   154

CHAPTER 10: The Golden Rule   169

CHAPTER 11: The Triumph of Selflessness   190

EPILOGUE: The Joy of Giving   207

Notes   211

Bibliography   225

Index   245

Acknowledgments   253

About the Author   254

Preface

THERE IS ALWAYS A RISK IN TAKING a book whose thesis is meant to be timeless and placing it in a present-day context. One runs into the danger of reducing a more universal argument to mere current events—and of no longer being understood once the headlines of today have faded from memory. And yet, since the publication of the original German edition in 2010, so many of the ideas explored in *Survival of the Nicest* have been affirmed that when my American publisher asked me to take stock, I gladly agreed.

This book is the fruit of decades of grappling with a particular version of Darwinism that has been gaining traction—and not only in the sciences—since the 1980s. In short, this perspective understands life as a genetic competition in which each individual tries to attain as many short-term advantages as possible. Quite often, however, it has only been through collective means that we have attained a social coexistence. And so, I wondered if, aggression and greed notwithstanding, an inclination toward cooperation in our nature hasn't also emerged, if a tendency toward fairness and generosity couldn't also be present in human nature. Numerous new studies, which you will find described in the pages that follow, have confirmed my suspicion.

The way our society has radically changed in recent years can be explained in light of this new and increasingly accepted take on the origins and nature of human beings. Those who believed that a society functions best when every person can follow the pursuit of profit with as few restrictions as possible in the free market were taught otherwise by the near collapse of the global financial system in September 2008. Perhaps the story of those dramatic days could be compared to the autumn when Berliners stormed the Wall: The Communists' conception of humanity at that time was finally destroyed by reality. Now the opposing—and just as naïve—ideology of the West has failed, too, after the egoism of the few has—as the following pages will prove— given way to the greater well-being for all.

What now? When the full brunt of the recession hit after the melt-down of multinational banks, observers predicted the disintegration— even the fall—of society. Those who faced uncertain economic survival, so the alarmists claimed, would only be mindful of their own welfare. A full stomach would come before ideals, to paraphrase Bertolt Brecht. And yet, that's not what has happened. The years since the col-lapse have shown to an impressive degree just how pronounced the human tendency toward cooperation is. For example, the Great Reces-sion in the U.S. in no way led to fewer people performing volunteer work; according to data from the Labor Department, the volunteer rate actually climbed during the years 2007–2009. An astonishing number of people have interpreted the collapse and Great Recession as a sign of not only economic but moral upheaval as well, spurring a long-overdue debate about social inequality that is now in full swing on both sides of the Atlantic.

People in Southern Europe have especially suffered—and continue to—because of the crisis. In Greece, a country particularly close to me since my wife is Greek, the unemployment rate has tripled since 2010, millions of families have fallen into poverty, and senior citizens can no longer pay their medical bills. Yet, here is a place where hardship has also resulted in people stepping up to take responsibility for each other in a way that hasn't been seen in decades. In today's Athens, you can find "Barter Bazaars," families cooking for their needy neighbors, and doctors treating their patients for free. Young people are returning from cities to the villages that their ancestors left and trying out new forms of communal living there. As I'll detail in the pages that follow, the argument for a cooperative nature in human beings anticipates pre-cisely this rediscovery of solidarity in crisis.

Moreover, the triumph of the internet in recent years—a realm more insulated from economic decline—has further enabled a culture of sharing to blossom. For many people, it has become routine to share their cars with their friends or even to open up their apartments to couch surfers. And the amounts of money that have been raised on crowdfunding platforms in order to realize a common project, or even to pay a complete stranger's college tuition, number in the high millions. Phenomena such as these are anything but surprising given how convenient this burgeoning network has made new forms of altruism.

How much trust must we summon? What measure of justice can we demand? How do we find the right balance of self-interest and altruism in our own lives and in society? Today, these questions are posed in a fundamentally new way—with an urgency that, even a few years ago, most people hardly would have expected—and they will accompany us for a long time to come.

In order to come closer to answering them, we need a realistic understanding of human nature. It is the only way that we can live happily as individuals, that society can flourish. This book tries to convey the necessary facts. If it can continue to be useful, I'll be happy.

Berlin, August 2014
Stefan Klein

Introduction

SOME POEMS ARE LIKE OLD FRIENDS. Companions through the years, they keep us fascinated even when we don't entirely understand them. That's what the following lines were like for me:

> On my wall hangs a Japanese woodcut
> Mask of an evil demon, lacquered in gold
> With empathy I see
> The swollen veins on its brow, suggesting
> How stressful it is to be angry.

I was seventeen when I read Bertolt Brecht's "The Mask of Evil" for the first time. Like so many young adults, I was angry at the world and longed for a better one. Of course I understood the poem's literal meaning; from my own experience I well knew the strength required to quarrel and the energy wasted in being angry! Worse than the unpleasant feeling itself is that it separates you from other people. Fury is a prison. Each object of our anger is one less person we can join forces with.

But Brecht's word for "angry," *böse*, designates more than just a feeling. It's a moral judgment as well, for it also means "evil." This is almost certainly what Brecht had in mind when he entitled his poem *Die Maske des Bösen*. He wrote it in 1942, when the Nazis' conquest of Europe was at its height and Hitler's troops were spreading terror from Norway to North Africa, from the Crimea to the Atlantic. But the angry young man I was when I first encountered his poem found this reading of the poem extremely irritating. How was it possible that people who exploited, injured, and killed others for their own profit could suffer from that behavior? Do Himmler and Hitler in the end deserve our pity?

Much later I understood that you could turn this idea around. If we remain free of malice and show ourselves to be fair and generous, it's possible that we do so not just out of fear of punishment or because it's been hammered into us by our upbringing. Humaneness in our dealings with others could perhaps also benefit us by raising our own sense of well-being. The ancient question of whether to worry about the happiness of others or only about your own would then automatically be answered: Both should concern you, because one can't exist without the other.

That thought was the germ of the present book. It sets out to contradict all the admonitions to proper behavior as well as the centuries-old philosophical principles that tell us we must resist our sweet, egocentric inclinations because that is what the bitter pill of moral duty requires. If our own well-being is so closely bound up with that of others, that would explain why so many people who chase after their own private happiness fail to find it: Perhaps these seekers of happiness have chosen the wrong goals.

More than twenty-five hundred years ago, Aristotle was already postulating that a happy life kept the welfare of others in view. But the philosopher had no way to prove his speculation, which is another reason why the idea took hold that moral action can only occur at the price of self-denial. But today we have empirical research that confirms Aristotle's conjecture. Humans who act for the benefit of others are as a rule more content and often more healthy and successful than contemporaries who think only of their own welfare. "One thing I know," Albert Schweitzer once confessed, "the only ones among you who will be truly happy are those who have sought and found how to serve."[1] In that spirit, this book is a continuation of my earlier work *The Science of Happiness*.[2]

Do altruists really get through life better? Everyday common sense rebels against the notion. If you give something away, you have less for yourself, while people who use their strength or spend money to achieve their own goals have an advantage—at first glance, anyway. A look at the natural world seems to urge us to hold on to what we have, for all animals, including humans, are competing for scarce resources. Those who have, survive; those who don't, perish.

But I intend to show that everyday common sense is mistaken. Our life together follows much more complicated rules than the law of the jungle. The following pages will explain some of the principles that actually govern our success or failure. A central discovery is that egocentrics do better only in the short term, but in the long run, it is mostly people who act for the welfare of others who get ahead. Of course "mostly" doesn't mean "always," and we will need to discuss when the one strategy or the other is called for.

If people who help others are more successful, evolution ought to favor such behavior. And this introduces a fascinating hypothesis: Might it be innate for us to care for others? Is there a gene for altruism?

The fact that the world is teeming with egocentrics is not an argument against the possibility of such a gene. Because of course, people are not programmed *solely* as selfless beings. It's even possible that our predisposition to look first to our own advantage is strongest. That would explain why mere exhortations and resolutions to be a better person are so ineffective. But the question is not whether a certain measure of egocentrism is unavoidably part of being human. Instead, it's a question of whether we possess other and less well-known innate impulses.

Humans are more conflicted in their motives than any other creature. What's more, we possess the unique freedom to act against our instincts. The spectrum of applications for our innate talents is enormous. For example, evolution constructed us as runners; that's why every healthy person is capable of running a marathon after the necessary training. But many of us use our cars even for the shortest errands and allow our leg muscles to atrophy. In the same way, we can cultivate or neglect our predisposition to altruism.

Nature, however, has a clever way to get us to do what she wants— she seduces us with good feeling. Sex is exciting and pleasurable because it serves reproduction. Our sensations of pleasure while eating, which are more effective than many would wish, aim to store up a layer of fat against lean times. In a very similar way, nature rewards us for fairness and helpfulness; it feels good to be generous. Brain research in fact shows that altruism activates the same synapses as eating a chocolate bar or having sex.

One is tempted to paraphrase Brecht: "How sad it is to be egocentric"—sad and dangerous as well. Not dangerous for one's fellow humans, since at least in developed societies, unbridled egocentrism is kept in check by laws and courts. But who is going to protect the egocentrics from themselves? Serious depression is on a frighteningly rapid rise in most countries, including Germany, where I live. Within a single year, the risk of young people becoming clinically depressed has more than tripled. And according to the World Health Organization, in another ten years, depression will be the most prevalent disease among women and second only to cardiovascular diseases among men. Many experts explain these frightening statistics by pointing to the dissolution of the traditional bonds among families, friends, and colleagues, which results in societies in which only the individual counts. What is certain is that a commitment to others can prevent morbid melancholy.[3]

Then what keeps us from caring more for others, if only for our own good? Whoever tries it soon realizes how often we second-guess our own wish to be generous. Although we may frequently feel an impulse to do something for others, we often suppress it. For altruism almost always seems riskier than acting exclusively for our own advantage.

For one thing, there is the fear of making ourselves look ridiculous. Generosity has a strange reputation in our society. We praise selfless human beings in public but remain cynical in private. We reserve our admiration for those who seem cool and strong-willed. Empathy, on the other hand, is considered a sign of weakness. The good judgment of those who occasionally put their own interests second is called into question; all too often one hears the word "do-gooder."

And so we are hopelessly ambivalent in the matter of selflessness. We want to believe in it but can't, and even if we could, we wouldn't admit it. But one thought seems not to occur to anyone—that someone's willing commitment to others could be a sign of strength.

Even deeper than the fear of derision is the completely rational fear of being exploited. For as long as people strive for their own advantage, some will take advantage of the goodwill of others. That has been the tragedy of every revolution begun by idealists.

And so this book will speak of giving and taking, of trust and betrayal, empathy and ruthlessness, love and hate. But the question will not be whether humans *are* good or bad. The greatest philosophers have already puzzled long and hard about that. What they have written sometimes sounds like a discussion about whether motion pictures as a whole are entertaining or disturbing—it depends, of course, on what movie you see. Nor will this book be about how we *should* behave. There are already plenty of convincing systems of moral philosophy, and the only question is why we so seldom follow them.

What I will try to explain instead is under what conditions humans are fair and generous—and when they are unscrupulous and egocentric. We must differentiate three questions. First, how is unselfishness possible at all? Second, what moves us to do things for others? And finally, why are some people so much more helpful than others?

The first section of this book will focus on the most clear-cut but by no means simplest form of living together: you and I. Our propensity to share will be explored, but also our propensity to cheat. For although cooperative action pays off, so does cheating, at least in the short term. But if the generous person who attributes good motives to others and is forgiving usually does better in the long run, how do we decide when we should be trusting and when it would be better to hold back? The demands such a decision places on our reason can be too great to sort out. The empathic system often comes to reason's assistance, for it functions quite differently from the usual strategic thinking. Whenever we encounter others in joy or pain, we mirror their feelings in our own head. As if the border between "you" and "me" were dissolved, the two brains then resonate together. Similar mechanisms assure that trust and mutual understanding occur.

The empathic system has an extraordinary number of facets. Contrary to what is often asserted, empathy alone makes us neither generous nor ready to help. Active help for others requires that we can feel what moves another person. And finally, neuroscientists have recently succeeded in providing impressive visual images of how friendship and love originate in our heads.

The focus of the second part of the book will be the community, beginning with a journey into the distant past: How did our ancestors learn to share with one another? This is still one of the greatest puzzles of evolutionary theory. Often enough, humans have been criticized as the cruelest of all creatures. But in fact, we are uniquely magnanimous. The most recent research shows that no other animal voluntarily hands over anything to another member of its species except for giving food to its own children. But all over the world, humans see to their nourishment collectively, and even little girls and boys spontaneously give presents to each other. There is much evidence that our ancestors first had to become the friendliest apes before they got the chance to become the smartest apes as well. We owe our intelligence to our willingness to give.

But we don't give indiscriminately. One of the strongest and most vitally important of all our needs is our desire for fairness. A community that does not ensure fairness among its members will fail sooner or later. Justice makes altruism possible, but the hunger for justice also brings with it envy and the desire for revenge. And these are not even the darkest aspects of selflessness; every group sticks together best when it is in competition with other groups. That's why exclusion and hatred of "the others" are the dark sisters of altruism. Thus humans owe their propensities to take care of others not only to their most noble but also to some of their ugliest characteristics. In this regard, modern research confirms a relationship portrayed in the myths of all ages, from the fallen angel Lucifer to Darth Vader: the figure of light transformed into a dark tyrant.

Can we live out altruism's good sides while avoiding the bad? The future of humankind depends largely on the answer to that question. As long as corporations, peoples, and nations pursue their own interests at the expense of the welfare of all, it will hardly be possible to protect the bases for life on our planet.

The history of humankind began with an altruistic revolution—our ancestors started to care for their fellows. Only together did they stand a chance in a world where food was growing scarce because of climate change. Today we find ourselves at a similar threshold: The challenge is to learn cooperation on a much larger scale. It is time for a second altruistic revolution.

There is good reason to be optimistic. Through digital networks, easy travel, and global trade, far-flung regions of the world are drawing closer together and cultures are merging at a breathtaking pace. In this book, I would like to show how this network of connections also shifts the things that drive our behavior. It costs us less and less to be selfless, while egocentrism grows more and more risky.

The future belongs to the altruists. We are born with the predispositions necessary to maintain ourselves in the world. But while we are familiar with the rationally justified pursuit of our own advantage, we are still uncertain about the impulses that lead us to seek our own happiness in the happiness of others. This book is an invitation to explore the friendly side of ourselves.

PART I

YOU AND I

CHAPTER 1

The Unexplained Friendliness
of the World

I have received in a Manchester Newspaper a rather good squib, showing that I have proved "might is right," & therefore that Napoleon is right & every cheating Tradesman is also right.

CHARLES DARWIN[1]

WESLEY AUTREY WAS WAITING FOR THE SUBWAY in New York City with his two little daughters when the young man next to him suddenly began to shake, went into convulsions, fell on his back, and began to wave his arms and legs like an overturned beetle. There were perhaps a hundred people on the crowded platform, but most of them looked the other way. Besides Autrey, only two women rushed over to help, but Autrey was faster. With great presence of mind he asked for a ballpoint pen and jammed it between the young man's teeth so he couldn't bite his tongue during the epileptic seizure. After a short time, the convulsions lessened and stopped, the young man got up, and Autrey thought he would be able to continue on his way.

A rumble and the glare of a headlight announced the arrival of the train. At that moment, the epileptic began to stagger again. He stumbled to the edge of the platform, lost his balance, and fell onto the tracks. Autrey asked one of the two women to look after his daughters and jumped down onto the track bed. The train was already rolling into the station, leaving Autrey not even a fraction of a second to think. He grabbed the fallen man and tried to heave him onto the platform, but the man was too heavy. So Autrey pulled him down between the rails and threw himself on top of him. The epileptic struggled, but Autrey pushed him down with all his might. When something cold touched his forehead, Autrey pressed his head into the other man's shoulder. Only two fingers of space separated his head from the train's undercarriage.

Five cars rolled over him. Then the train came to a halt and Autrey could hear his daughters screaming. When rescue workers finally freed the two men from beneath the train, there was a smear of grease on Autrey's cap. The paramedics discovered only a few bruises on the epileptic. Autrey himself refused medical attention. In his opinion, he hadn't done anything special even though he knew he had risked his life. "I saw the man and he needed help . . . I just felt that I had to do something."[2]

If you imagine Autrey as a taciturn and straitlaced fighter for justice and decency—a western hero in the mold of Gary Cooper—you would be mistaken. And he is nothing like the cliché of the pale martyrs eager to make a show of their readiness to sacrifice themselves for others. Wesley Autrey has an athletic build, and if you met him in his Harlem neighborhood, with his warm-up suit and a baseball cap turned backward, you might mistake him for a rapper. Only a few gray hairs in his beard hinted at his age of fifty-one.

His actions at the 137th Street stop on that day, January 2, 2007, made him a national hero. He was invited to the White House and interviewed on talk shows, where he spoke with such animated and articulate ease that he seemed used to such attention. In reality, he earned his living as a foreman on construction sites and had earlier worked three years as a Postal Clerk 3rd Class in the United States Navy.[3] But if anyone was awkward in his interviews, it was prominent interviewers like David Letterman, who made lame jokes to distract attention from the fact that he couldn't match the eloquence of his guest. Autrey proved to be a much cooler guy.

The media and politicians celebrated him as an example, and when Autrey now entered the station at 137th Street, people constantly wanted to touch him as if to assure themselves that he really was a flesh-and-blood human being.

But no one seemed to notice how disturbing Autrey's heroic deed actually was. What brings a father to risk his life for a stranger in the presence of his two children, aged four and six? How can a person decide in the space of a few seconds to risk his own life to save another?

The Hero Next Door

Millions of TV viewers may have admired Autrey, but what he did represented a real challenge to science. According to traditional scientific explanations, the events beneath 137th Street should not have taken place. The last few decades of research in behavioral science have produced an image of humans as deeply selfish beings. Biologists have viewed us as programmed for maximal reproductive success. Evolutionary psychologists say we are hardwired to seek status. And most economists (probably the most influential of all social scientists) understand human activity as a search for affluence and comfort. All the disciplines have been unanimous in their assumption that everyone is looking out for number one and altruism is an illusion.

But researchers today clearly recognize the difficulties raised by such a conclusion. After all, both men and other animals live and work together in peace. The fish called the bluestreak cleaner wrasse (*Labroides dimidiatus*) swims into the mouths of potential predators, like groupers and moray eels, which could swallow it in an instant. Instead, these predators allow it to stay and eat the parasites that have collected there.

Ants, bees, wasps, and termites live in colonies of millions of individuals, proving that cooperation in large groups can be spectacularly successful.[4] For how seemingly insignificant each individual insect is, their communities are all the more impressive. It is estimated that in the tropics, termites alone constitute half the animal biomass; that is, these social insects taken together weigh as much as all the other animals inhabiting tropical Africa, South Asia, and Central and South America. Not even the spread of our own species has reached such proportions. The weight of more than seven billion human beings in the world is equal only to the weight of all other vertebrates combined.

Yet *Homo sapiens* rules the entire planet and has formed globe-spanning organizations.

All of this would be incomprehensible if individuals had only their own interests in view. And so behavioral scientists worked for decades to explain how communities are possible if every action must have personal advantage as its reward.

Yet how could they possibly explain the fact that again and again, people like Wesley Autrey selflessly come to the aid of others, even at the risk of their own lives? Heroes may be few and far between, but can we simply dismiss them as exceptions?

After all, tens of thousands of people risked their lives to save Jews from the death camps during the Second World War. And a great number of my fellow citizens are willing to suffer pain for others: More than three million Germans have registered to donate bone marrow for leukemia patients unknown to them. In the United States, there are popular websites where volunteers can donate one of their kidneys for transplantation into a stranger.[5]

Less spectacular but perhaps even more important for our coexistence are the innumerable situations in daily life that do not conform to the image of an always-egocentric human race. Why, for example, do we leave a tip even when we know we will never return to the restaurant? Why do we rush to stop a child we see running into the street? It's also difficult to see the advantage of caring for a bedridden relative for years, giving an anonymous donation to help earthquake victims we don't know, or volunteering our precious free time for a good cause, as more than sixty million Americans do each year. And my own Germany would certainly be a different place today if, roughly twenty-five years ago, first hundreds and later tens of thousands of East Germans had not braved possible repercussions from the Stasi to turn out for the Monday demonstrations that led to the fall of the Berlin Wall.

And while the percentage of Americans involved in traditional forms of civic engagement is stable, brand-new forms of cooperation and selflessness are blossoming on the Internet. People worldwide donate their expertise for free, enabling ten million articles to be put up on Wikipedia seemingly overnight, as well as the free open-source programs that challenge the dominance of firms like Microsoft.

In the case of many scientific puzzles, we can easily accept the fact that researchers have not yet been able to solve them. But the fact that altruistic acts are difficult to explain and yet ubiquitous raises fundamental questions about our very concept of ourselves. How selfishly—and how selflessly—are people capable of behaving? Under what circumstances do they put their own interests second? How can engagement on behalf of others be encouraged?

We often complain about the selfishness of our contemporaries. But perhaps human kindness is like the air: We constantly move within it and can easily forget entirely that it is there. Only when we're deprived of it do we realize what we're missing. The person who has a meal in a restaurant with friendly service and doesn't leave a tip is universally considered a miserly boor.

The Mother Teresa Problem

Let us first be clear about what egocentrism and altruism are. In everyday speech, these words usually have a moralistic connotation, depending on the perceived motives behind any given action. We call people egocentrics when we believe they only think of their own advantage. We call them altruists when we believe they only want the best for others and nothing at all for themselves. If you give a beggar the shirt off your back because it makes you feel good, you would not be an altruist by this definition. For even if you're left naked, you didn't take off your shirt for the beggar but to make yourself feel good.

According to these definitions, people who consider humans as purely egocentric are almost always right on a superficial level. For in the end, we can assume that all benevolent people derive satisfaction, pride, or some other uplifting feeling from their good deeds, and in most cases enjoy the admiration of their fellow men as well. And don't many contemporaries engaged in helping others say that it is precisely this engagement that has given meaning to their lives? People who act selflessly thus seem in some subtle way to be even more egocentric than others.

But that is too simplistic a view. First of all, it still doesn't answer the question of *why* we feel good when we do things for others. And second, if people are in high spirits after their altruistic acts, that doesn't necessarily mean they acted altruistically only in order to feel good. The common understanding of egocentrism is inadequate because we can never know the true motivations of others. But that does not mean all altruists are being dishonest. People often don't know exactly why they do or don't do something.

That's why it is so hard to contradict the argument that altruists help others only for their own satisfaction. Cynics like to assert that even Mother Teresa did not act selflessly. According to them, the nun who collected the dying from the streets of Calcutta, washed the wounds of lepers, and lived voluntarily in a slum, simply felt good taking care of the poorest of the poor.[6]

We happen to know quite a lot about the psyche of Mother Teresa, since for years she relentlessly interrogated her own interior life. Not long ago, her diaries and private correspondence were made public, and those documents show how pitilessly the Nobel laureate questioned her own motives and actions.[7] For many years, this woman who had devoted herself to the service of Christ felt abandoned by God and doubted his very existence. She was even more suspicious of her own feelings. She wrote that it was ice-cold inside her. So even these intimate writings do not shed light on what motivated Mother Teresa—obviously she didn't know the reason herself.

However, there are cases in which we can almost certainly exclude any hope of praise or a good feeling. Did Wesley Autrey risk his life for a stranger so he could be celebrated as a hero? The speed with which he had to act makes that highly unlikely. In the split second he had to decide whether to jump down onto the tracks, Autrey had no chance to imagine how he would feel if he were successful. And even if he had had time for such thoughts, the possibility of getting to shake hands with the president wouldn't have been much motivation in the face of the high likelihood of being killed.

It's just as unlikely that honor, a sense of duty, or his conscience played much of a role at the crucial moment. In that terrible second there was simply no time for such complex considerations. And so he must have risked his life for reasons beyond his conscious control. Even in the common understanding of the word, Autrey acted completely altruistically.

But are the acts of people who, after careful thought, decided to harbor persecuted Jews to save them from the death camps less admirable? Or are they perhaps even more so?

A Dozen Good Deeds a Day

It doesn't get us very far to adduce this or that hazy motive as a measure of selfishness or altruism. It's more fruitful to look at the risks and rewards behind any given deed. Every action has costs and—hopefully—benefits. And to ask who bears the costs and who reaps the benefits reveals immediately whether someone is acting selfishly or altruistically. A selfish person enjoys a benefit that others pay for. The behavior of a thief, for instance, is extremely selfish. An altruist, on the other hand, accepts costs in order to create benefits for others—by giving something away, for example, without expecting anything in return.

This definition is also made use of in behavioral research. It is simple and useful because one can look at costs and benefits, unlike motivations that often remain unknown. To be sure, an act of altruism does not necessarily require a cost. It suffices to take a risk for another person. Wesley Autrey emerged unharmed from under the train that had rolled over him. But since it could have been otherwise, Autrey's deed was altruistic.

Nor does altruism demand that we sacrifice ourselves for a needy person. Even if we just forego some small advantage as a favor to others, we are acting altruistically, according to the scientific definition. The beneficiary does not need to be a specific person, either. We often act altruistically for the welfare of a group or even for some abstract principle—fairness, for example. And even if benefactors are members of the group they help and thus share in the benefit themselves—for example, by voluntarily bringing their trash to the recycling center—they are acting selflessly to the extent that their costs (the time they spend doing it) outweigh their own share of the benefit (an environment that is a minuscule amount less polluted).

So whether we return the excess change a checkout clerk has accidentally given us or accept a package for a neighbor or agree to serve on the PTA, we bear the cost while others reap the benefit. Seen in this light, our everyday existence is full of altruistic actions small and—sometimes—large. Boy scouts are supposed to perform a good deed every day, but in fact most people make selfless choices dozens of times every day.

Is Sex the Answer?

Whoever dismisses the examples above as banal is overlooking how hard they are to explain. Even the skeptics who don't believe there is such a thing as a selfless action admit this difficulty. But they defend themselves by saying that costs and benefits are not so easy to define clearly. Everyone knows that humans go at things in indirect ways. The father so concerned with the good of the entire student body that he joins the PTA could actually be trying to ingratiate himself with the principal to give his own child an advantage. "Scratch an 'altruist' and watch a 'hypocrite' bleed," scoffed Michael T. Ghiselin.[8]

So what constitutes a cost or a benefit? It's easy to see why probably the most frequent way to measure them so often fails to offer insight into altruism. Usually we use the words "give" and "take" with reference to possessions or time, both of which can be converted from one to the other: How much time will it cost to acquire a certain item? Asked for a favor, most people will calculate it in the same way. To a great extent, both traditional economics and the politics that depend on economic theories rest on a purely materialistic understanding of costs and benefits.

But if we try to solve the riddle of altruism on this basis, we become unavoidably tangled in absurdities. If all anyone cared about was getting rich in the easiest possible way, then not a single child would have been born since the advent of effective contraceptives. Obviously, the cries of a baby represent a benefit to its parents not expressible in dollars and cents, a benefit for which they are prepared to spend absurdly long hours in the nursery and incur many thousands of dollars per year in expenses. Of course, no one would claim that people begin a family out of altruism.

Evolutionary biologists have come up with a sensible measure of costs and benefits: Like all organisms, humans are programmed to pass on their own genes. In this light, everything that serves that goal in the long run is a benefit. The individual who does something that diminishes his chances for reproduction, however, incurs a cost. An egocentric logically increases his own reproductive chances by diminishing those of others. Altruists do the exact opposite.

The evolutionary advantage explains things much more plausibly than the purely materialistic bookkeeping of the economists—for example, humans' lively interest in sex, for which some are even prepared to pay money. Nor is it surprising that people battle for status, given the fact that prestige can only enhance one's chances of reproductive success. Status symbols, after all, attract sexual partners like a light attracts moths. Yet Porsche owners don't have more or healthier children than other men; our evolutionary programming is a remnant of the past. It was developed at a time when status could in fact increase the number of descendants one produced. Yet it determines our behavior to this very day.

The competition for the chance to reproduce thus explains what can appear at first glance to be strange behavior. Moreover, the approach adopted by evolutionary biologists affords us a glimpse into the future. If a particular innate behavior increases the number of descendants, then it will automatically spread into the next generation, because parents with the corresponding innate predisposition will produce an above-average number of children who will pass on their genes.

Since the advantage adds up from generation to generation, it may be quite small to begin with. But whoever has genes that are even a tiny bit more advantageous than others is like a saver who invested a single penny in one of the new banks being founded at the time of Christopher Columbus: Even after adjusting for inflation, that saver's descendants would have more than ten million euros today (if only there hadn't been any currency reforms). In evolution, a genetic advantage pays dividends over an immensely longer period of time.

The Decline of the Noble Warrior

And here is the strongest argument that the theory of evolution provides to the skeptics who deny that there is any such thing as altruism: If selflessness means acting against one's own best biological interest, how can such an evolutionary handicap continue to exist over the long term? Altruists assume costs in the form of reduced chances of reproduction; the benefits accrue to others who are thereby able to raise more children. This problem was already recognized by Charles Darwin, the father of modern evolutionary theory. In a world of pitiless competition for resources, there seemed to him little room for generosity. Darwin illustrated this bitter insight with the example of a warrior people: "He who was ready to sacrifice his life, as many a savage has been, rather than betray his comrades, would often leave no offspring to inherit his noble nature. The bravest men, who were always willing to come to the front in war, and who freely risked their lives for others, would on an average perish in larger numbers than other men."[9] After a few generations, the altruists would have died out.

From this perspective, a predisposition for selfless action would never have become established, and not just in such dramatic cases as a soldier's self-sacrifice on the battlefield. As the skeptics maintain, even minimal disadvantages count in the course of evolution. If you take care of the elderly, you have less time to look for a mate. If you donate money anonymously or even return the excess change a cashier has mistakenly given you, you have less money for your own children. Wesley Autrey even left his daughters behind to risk his life for a stranger.

But with this argument, the skeptics have maneuvered themselves into a corner. If all that counts is one's immediate biological advantage, then there ought not to be any such things as care for the aged or charitable donations or a person like Wesley Autrey. Is reality out of whack if it doesn't fit the theory?

But despite such obvious contradictions, to this day it remains enormously popular to assert (with Darwin as backup) that we're all just out for ourselves. Whoever needs to justify ruthless or even immoral behavior can use this argument to give it a whiff of scientific legitimacy. Such rhetoric is as old as the theory of evolution itself. The English philosopher Herbert Spencer, a convinced liberal and contemporary of Darwin's, coined the phrase "survival of the fittest." In books that sold hundreds of thousands of copies, Spencer was the first to attempt to derive rules for living together from the theory of evolution. He wrote, for instance, that it was absurd and even wrong to worry about the weak, for "the whole effort of nature is to get rid of such—to clear the world of them, and make room for better."[10]

Redefined as the "struggle for existence," similar ideas would recur in Hitler's *Mein Kampf*. Spencer was also widely read in the world of commerce. John D. Rockefeller, by the early twentieth century the richest man of all time, invoked natural egocentrism while bringing all of North America under the sway of his Standard Oil Company and then defending his monopoly with all the means at his disposal. According to him, the growth of the company was a manifestation of the survival of the fittest, "merely the working-out of a law of nature and a law of God."[11] Michael Douglas as the speculator Gordon Gekko in the 1987 film *Wall Street* famously summarized the spirit of the past few decades in the lines "Greed . . . is good. Greed is right." He too had recourse to biology: "Greed . . . captures the essence of the evolutionary spirit."[12]

Darwin's Dilemma

The most frequent objection to such a simplistic Darwinian view is that fairness and willingness to help others became instilled in us against our human nature. Morality, it is said, is a purely cultural achievement. This is the argument made by most social scientists and also a surprising number of evolutionary biologists.[13] But this thesis doesn't get us anywhere, for it fails to explain why people in a crisis follow the rules they were once taught. An egocentric would simply ignore the precepts of her moral education if doing so promised to benefit her.

Now, we could surmise that we've had moral values permanently burned into our childhood brains by some mysterious mechanism, but that still doesn't solve the basic problem. If we explain morality as a product of culture, we simply shift the riddle into the past. After all, something must have led our ancestors to teach their children a certain measure of altruism against their biologic self-interest.

And then this behavior, detrimental to themselves, had to be perpetuated over many generations. So Darwin's depressing argument cannot be challenged by recourse to culture. The warriors prepared to sacrifice themselves have reduced their own chances to reproduce, and whether their magnanimity was innate or acquired makes no difference.

Charles Darwin found no solution to this dilemma. In 1902, the Russian anarchist, writer, and universal man of learning Prince Kropotkin proved convincingly that many animals display the same mutual assistance observed in humans and thereby raise the biologic fitness of their species. But he couldn't disprove Darwin's assertion that the helpful place themselves at a disadvantage vis-à-vis the egocentrics *within* the group. The work of the Russian nobleman was soon forgotten, not least because his radical leftist politics made him an outsider in the world of traditional science.[14]

Darwin's followers in Western Europe and America, meanwhile, had only one plausible explanation. Altruism, they said, could be explained, but only among blood relatives. Thus, if a father foregoes something in favor of his daughter, that promotes the perpetuation of his genes. If the father's care for his family is innate, that predisposition will be passed on. And genes that favor altruism within a family can propagate not just by direct inheritance, for an aunt who devotes herself to her nephew also has a biological advantage. After all, according to Mendelian law, a fourth of his genes will also be hers. If she looks after two of her nephews, she has done as much to perpetuate her genotype as if she had cared for a child of her own.

It was the English geneticist J. B. S. Haldane who made this observation. He was once asked if he would jump into an icy river to save his brother from drowning. "No, but I would to save two brothers or eight cousins."[15] Statistically, that would ensure the continued existence of all his genes.

The adherents of such ideas called themselves "sociobiologists" and succeeded, for one thing, in explaining altruism in ant colonies. As a rule, it turns out that the closer these social insects are related, the more likely they are to sacrifice themselves for one another. From the knowledge of the number of shared genes, it is possible to calculate how much an ant—or a bee or a wasp—will do for another of his species, exactly as Haldane calculated whom he would save from drowning. Inspired by their experimental successes, the sociobiologists made a serious attempt to do what for Haldane was only a witty rejoinder: to apply their theories to human beings.

Regrettably, they thereby missed the more interesting question. A group of humans is not an ant colony whose members are all related to one another. If you want to understand our communal life, you have to explain why we support and share things with people *outside* our family. But rather than address this problem, the sociobiologists discussed only what they could solve with their approach: altruism *within* a family.

What Groucho Recommends

To escape this dilemma, the sociobiologists posited that altruism could not exist outside the family. If people did something for someone else, they were only speculating that they would get something in return. And if they seemed to be acting in a fair and benevolent way, it was only to camouflage the pursuit of their own advantage. Wasn't it Groucho Marx who said, "The secret of success is honesty and fair dealing. If you can fake those, you've got it made"?

Richard Dawkins' *The Selfish Gene* appeared in 1976 and made its author the world's most famous sociobiologist. The brilliantly written text not only became a bestseller but also influenced an entire generation of biologists and behavioral scientists. Dawkins' pronouncements on his trademark "evolutionary psychology" continue to be quoted in both the scientific literature and popular self-help books. (Among Dawkins' fans was Jeffrey Skilling, the Enron CEO responsible for the biggest accounting fraud in history. He is said to have called *The Selfish Gene* his favorite book and the main source of his inspiration.[16])

Dawkins, who went on to become a vehement critic of religion, found stronger words to express the sober formulations of other colleagues: "We are survival machines—robot vehicles blindly programmed to preserve the selfish molecules known as genes." And since "like successful Chicago gangsters our genes have survived . . . in a highly competitive world," one cannot expect selflessness from gangsters or from us, for "this gene selfishness will usually give rise to selfishness in individual behaviour."[17]

But the life story of the intellectual godfather of sociobiology calls such harsh sentences into question. J. B. S. Haldane, who joked that he wouldn't jump into an icy river to save a brother but would to save two, was actually intensely concerned with the welfare of his fellow humans. A committed Marxist, he served as editor-in-chief of *The Daily Worker*, which didn't exactly further his career as a scientist. But his editorial work did not keep Haldane from becoming one of the most famous biologists of his time and, in the end, a member of the Royal Society. He supported the Republicans in the Spanish Civil War and wrote more than a dozen books in which he propounded his ideas about a better society to a broad readership. He even wrote a children's book. Horrified by Stalin's reign of terror, he later left the Communist Party, and in 1950 he left England as well in order to contribute to the development of an impoverished but newly independent India. There he became a vegetarian. His altruism was still audible in his last will and testament, in which he left his body to a provincial medical school. "My body has been used for both purposes during my lifetime and after my death, whether I continue to exist or not, I shall have no further use for it, and desire that it shall be used by others. Its refrigeration, if this is possible, should be a first charge on my estate."[18]

The Unrecognized Altruist

It is common to associate Charles Darwin's name with the principle of "eat or be eaten" and the ruthless struggle for survival, but in reality, what goes under the name of "Darwinism" in the popular imagination and even among scientists is a distortion of what the great English biologist taught. Even during Darwin's lifetime, the theory of evolution was appropriated for all sorts of political goals, which made Darwin himself extremely unhappy. The quotation at the head of this chapter is from a letter he wrote in 1860 to a friend, the geologist Charles Lyell. In sarcastic language, Darwin complains how the *Manchester Guardian* misinterpreted his just-published *Origin of Species*. Even worse, after his death some of Darwin's central ideas were forgotten while others weren't taken seriously.

While often portrayed as the prophet of egocentrism, Darwin was actually an extremely sympathetic person. Once while strolling past a wall in a coastal town in Brazil during his voyage on the HMS *Beagle*, he heard a groan of misery. Somewhere behind the wall, he surmised, a slave was being tortured, and Darwin was helpless to do anything about it. He couldn't even protest. This recollection was still vivid many years later. Later he would write, "I thank God, I shall never again visit a slave-country. To this day, if I hear a distant scream, it recalls with painful vividness my feelings, when passing [that] house near Pernambuco."[19]

In the southern English village of Downe, where he lived from his return in 1842 until his death, Darwin founded a Friendly Society that looked after the welfare of impoverished agricultural workers. Darwin could not see even animals suffering without being moved, his son reported. He returned from a walk "pale and faint from having seen a horse ill-used."[20] And when he heard that a farmer had let some sheep starve to death, he collected evidence and brought the case before a magistrate, thereby making himself unpopular in the village.

How was Darwin able to reconcile his altruism with his scientific discoveries? He was much too thorough a scientist and possessed too much life experience to accept as truth his own scenario of the noble warriors dying out. Even during the voyage of the *Beagle*, he noted in his diary events that were not compatible with a human nature that was purely selfish.

In Tierra del Fuego, the twenty-one-year-old Darwin encountered a people who seemed as strange as any he had ever met. He described his first encounter with a "savage" thus: "I could not have believed how wide was the difference between savage and civilized man: It is greater than between a wild and domesticated animal, inasmuch as in man, there is a greater power of improvement."[21] The inhabitants of Tierra del Fuego often went hungry and waged war for scarce resources. (Darwin planted vegetable gardens to at least somewhat improve their diet.)

And yet these men, who seemed to Darwin at first more like animals, possessed an innate sense of fairness. The young scientist discovered this when a small flotilla of canoes approached the *Beagle* in February 1834. "I gave one man a large nail (a most valuable present) without making any signs for a return; but he immediately picked out two fish, and handed them up on the point of his spear." Among themselves as well, the Tierra del Fuegans were willing to forego things voluntarily: "If any present was designed for one canoe, and it fell near another, it was invariably given to the right owner."[22]

This behavior had little to do with the struggle for survival. It seemed hardly imaginable that they had learned their sense of fairness from another culture, for the Tierra del Fuegans lived completely isolated on their island. So where did their morality come from?

This question must have haunted Darwin for almost half a century, for he returned to it four decades after his voyage. He devoted several chapters to it in the work of his old age, *The Descent of Man*. There he makes the bold claim that in the course of evolution, intellectual abilities and predilections develop in the same way as do physical characteristics. Thus many animals have an innate "social instinct" that makes them seek companionship and feel sympathy for other members of their species. In such intellectually advanced beings as humans, this instinct leads "inevitably" to an innate sense of fairness and morality.[23] And that is why altruistic behavior does not disappear in the course of generations. The innate propensity for community sometimes forces humans to act selflessly.

Darwin could not have foreseen the spiritual revolution to which these thoughts would later lead. He also still had no clear idea of what constituted this social instinct and how it could produce fairness and magnanimity. He speculated that primitive humans had developed virtues because they yearned for praise. Since he provided no further explanations, Darwin's followers were able to dismiss his idea that people naturally care for each other as a mere aberration of the master.

In actuality, however, Darwin was more than a century ahead of his time with his idea. Today we are only beginning to understand how much our thoughts and actions are marked by altruistic impulses and to discover their origin and how they benefit us in the long term. A new image of humankind is emerging that shows a much more friendly *Homo sapiens* than seen before. These insights are going to change the rules of our lives together.

CHAPTER 2

Give and Take

I charge you not to be one self but rather many selves, the householder and the homeless, the ploughman and the sparrow that picks the grain ere it slumber in the earth, the giver who gives in gratitude, and the receiver who receives in pride and recognition.

KHALIL GIBRAN[1]

WHY IS IT OFTEN SO HARD TO join forces with others, even when it would be to the advantage of everyone concerned? The Enlightenment philosopher Jean-Jacques Rousseau illustrated with a little parable how difficult human cooperation is:[2] Ages ago, some hunters set out to kill a stag together. On their way, however, a hare ran across their path. Each hunter could shoot the small animal and keep it for himself, but then their larger prey would be scared off.

Would they shoot? If you find Rousseau's stag hunt too far-fetched, just think of the quarrels in a normal relationship. A man and a woman—let's call them Adam and Eve—are married and both work full-time. They both love to cook and eat, but neither likes having to buy groceries. Of course, it would be only half as onerous if they shared the shopping. However, it would be even more pleasant for Adam if his wife went to the supermarket, the butcher, and the vegetable stand by herself. So he doesn't show up for their joint shopping trip. But Eve is also tempted to play hooky, surmising that Adam's longing for fresh vegetables and a nice chop will overcome his laziness. Since they both know they can't depend on each other, they end up with an unfortunate outcome—neither goes shopping and they spend another grumpy evening eating frozen pizza.

From taking out the trash to looking after the children, every household is full of such potential conflicts. Only two completely selfless individuals would take care of all the chores without arguments or annoyances. The rest of us struggle with the constant temptation to shirk our part of the responsibility. In the same way, each of Rousseau's deer hunters knows the others are tempted to shoot the hare. And before someone else gets the first shot in and they have to go home empty-handed, they will all shoot at the hare even though they know they're foregoing the larger prey by doing so.

Rousseau thus recognized a further obstruction to cooperation. While according to Darwin, the noble, self-sacrificing warrior must necessarily lose out in the evolutionary long run, the stag hunters would increase their reproductive chances if they all worked together. And yet among them as well, it is selfishness that wins out—with the result that everyone loses. For even the hunter who shoots soonest and kills the hare has to be satisfied with a comparatively meager supper.

In our couple, on the other hand, the partner who refuses to do his part may be lucky enough to get away with it. For if Eve knows that nothing in the world can get Adam to go to the supermarket, but for her, shopping is always more bearable than having to eat frozen pizza, she'll grit her teeth and buy the groceries. So Adam just has to be determined enough and he can completely avoid the unpleasant chore and still have a good supper. If Eve stands on principle and leaves the refrigerator empty, she'll be cutting off her nose to spite her face. The converse is equally true: If Eve can persuade Adam that she will absolutely not do the shopping, he will do it. Such are the raw materials for a war of nerves.

Naturally, each will accuse the other of having a character flaw and being unable to sustain the relationship. (After all, both have cogent reasons for their behavior. Doesn't he have much more stress at work than she does? Didn't she schlep bags of groceries by herself the last three times?) But actually, there's no reason for them to tell each other they need a therapist. They're both unwittingly following an implacable logic: If you want the best outcome for yourself, you have to be prepared to call the other person's bluff. And if you're unwilling to run that risk, you're in danger of being exploited.

Life as a Game

If you stacked up a single copy of all the novels, plays, poems, law books, and psychological studies that deal with peoples' difficulties in living together, the resulting pile would reach into outer space and touch the moon. That's how much we desire community and love and how greatly we fear betrayal.[3]

It makes the contribution of John von Neumann all the more impressive. As incredible as it sounds, the Hungarian mathematician managed to find a formula to express the basic theme of all conflicts between humans and even between other animals. One could say that he discovered a common denominator for the content of all world literature.

Until 1928, when von Neumann began to think about human relationships, people had seen the power of our passions as the essential cause of our quarrels. After all, it's obvious that they prevent us from finding reasonable compromise, the best way to solve any disagreement. But von Neumann realized that the path to this rational solution is frequently blocked—and not only because strong emotions prevent us from keeping a cool head. While fury and greed, aggression and jealousy can certainly hinder compromise, they also keep us from recognizing our real motivations. According to von Neumann, even very emotionally charged events such as sexual seduction or marital strife involve strategic calculation. This insight was visionary at the time. Based on it, von Neumann's followers would discover a way to motivate even confirmed egocentrics to cooperate with others.

There is no doubt that John von Neumann, born in 1903, was one of the most brilliant minds of his generation. (In fact, he considered himself the most brilliant human being of all time.) He is said to have amused visitors at his parents' house in Budapest, even before he started school, by telling jokes in ancient Greek and reciting entire pages of the telephone book from memory. By the age of twenty-three, he was teaching at the University of Berlin, where he not only solved several central problems of mathematical logic but also supplied the mathematical foundation for quantum physics, which was still in its infancy. A mere two years later, he published an article with the bland-sounding title "Theory of Social Games."

Actually, von Neumann understood all of life as a game. If you wanted to succeed at life, you had to find the best possible moves. And as is true in chess or poker, what's important is to keep one step ahead of your opponent. For only players who take into account others' possible next moves before deciding their own can make an optimal choice.

"Optimal" in game theory means obtaining the best outcome for oneself. In the case of the stag hunt described above, each participant must reckon with the treachery of the others, so the solution must be to shoot at the hare first yourself. Thus reason opposes cooperation and the idea of feasting on the stag together is an illusion. And although it would be best for our couple to do the shopping together, Adam and Eve have no hope of a well-stocked refrigerator, at least not as long as both act in a purely logical way. Whichever one announces a willingness to do the shopping must be prepared to have the other goof off. "It is just as foolish to complain that people are selfish and treacherous as it is to complain that the magnetic field does not increase unless the electric field has a curl," von Neumann said. "Both are laws of nature."[4]

Game theory gained currency because it could be applied to every conceivable problem of human life together. It can be applied to the daily chaos of commuter traffic as well as the friction in Adam and Eve's marriage. It's clear that every commuter would get home faster if they all used the bus instead of clogging the roads with cars. But there is no escape from the current situation. If just a few individuals use public transportation, it would take them even longer to get to work, since they'd have to wait for the bus and it would then get caught in the traffic jam too. And even if everyone switched to the bus, some would immediately be tempted to return to their cars because the roads would be invitingly empty. As inevitably as a ball always rolls to the lowest point in a bowl, people's lives together are always in a state of balance, and all too often, it is a balance of terror. That is the bitter lesson of game theory.

With its help, sociologists can also explain how societies hold together or fall apart and why the common good is such a fragile construct. Game theory tells economists why employees are unable to command higher salaries and companies to command higher prices, although it would be in the best interests of both to do so. And biologists use the same approach to investigate struggle and symbiosis in nature.[5] Thus game theory serves to unite all fields that deal with relationships among creatures.

John von Neumann also left his mark on politics—with disastrous consequences. After immigrating to the United States in 1930, he was one of the lead scientists in the development of the atom bomb in the desert of Los Alamos and planned its deployment against a Japanese Empire that was already as good as defeated. In the Library of Congress, one can see the piece of paper on which von Neumann wrote down the targets: Kyoto, Hiroshima, Yokohama, and Kokura. Thick fog saved Kokura from destruction; the second bomb was dropped on Nagasaki instead.

After the war, von Neumann the cynic (the model for Peter Sellers' character Dr. Strangelove in Stanley Kubrick's black comedy of the same name) had half a dozen contracts as consultant for the United States Army, the CIA, and the Truman Administration. His recommendations influenced American military strategy for decades. At a time when the superpowers threatened each other with ICBMs and any accommodation seemed impossible, von Neumann took his game theory literally. Since he considered compassion and a sense of justice to be illusions that humanity would abandon as soon as they became inconvenient, he pressed Washington to make a first strike in 1950. He said that nuclear war was inevitable in any case, so it would be better to carry out the first strike oneself. In an interview with *Life* magazine, he declared, "If you say why not bomb [Soviets] tomorrow, I say why not today? If you say today at 5 o'clock, I say why not 1 o'clock?"[6]

The Logic of Refusal

Doubtless, American strategists would have gained an advantage if they had bombed the Soviet Union shortly after World War II. So why didn't they? Why was the world spared a nuclear war? One could just as well ask why only 40 percent of all marriages in Germany end up in divorce court. According to the game theory of von Neumann (who remained faithful to his second wife Klari until the end of his life), any marriage between husband and wife ought to be basically condemned to failure.

The answer is this: Although the mathematician from Budapest had discovered an important truth, it wasn't the whole truth. One of the first to grasp this was John Nash. Until he developed paranoid schizophrenia, Nash was John von Neumann's colleague at Princeton—moviegoers will remember him as the hero of the film *A Beautiful Mind*. Decades later, he succeeded in taming the voices in his head, and he was awarded the Nobel Prize in 1994. Nash discovered that *every* strategic problem had at least one solution from which no player can deviate without disadvantage to herself, although the solution is usually not optimal. For example, our commuters stuck in rush-hour traffic find themselves caught in such a Nash equilibrium: Whoever tries taking the bus will have to put up with an even longer commute.

If Eve finds shopping more bearable than an empty refrigerator, she has no choice but to set off for the store and forget about fairness—if she can. To her, eating frozen pizza is a disadvantage; for her husband, going to the supermarket is more of a disadvantage. Thus Eve has a completely selfish motive to do something for her partner. (The same is true for Adam too, of course.)

But what if both of them are yearning for a steak? Then there is another solution that will escalate the war of nerves and is probably the most frequent in everyday life: Try all possible maneuvers to force the other to do what you don't want to, but leave yourself the final option of going shopping if worse comes to worst.

Seldom has this strategy been so graphically illustrated as in the scene from *Rebel Without a Cause* in which Jim Stark (played by James

Dean) and his rival are playing chicken in stolen cars. They race toward a cliff to win the heart of the lovely Judy, and the loser is the one who jumps out first. (Things go wrong, however. Jim jumps out at the last minute, but his rival, Buzz, catches his jacket sleeve on the door handle and plunges to his death.)

Thus to solve their problem, Adam and Eve have a total of three Nash equilibriums to choose from: (1) she always goes shopping since he refuses to go, (2) he always goes since she refuses, and (3) both threaten not to go, but will go if worse comes to worst. The last option is called a "mixed" strategy. It only works if both of them occasionally stick by their refusal to go. If they don't, the other partner will soon learn that the refusal isn't serious. With a little mathematics, one can even calculate how often they have to be stubborn. The resulting number depends on their personal preferences and will be higher the more a partner finds shopping unbearable and an empty refrigerator acceptable. If Adam and Eve continue with the mixed strategy long enough, both will learn how little they can depend on each other and will resign themselves to go to the store.

So it is that, surprisingly enough, a strategic dose of refusal can stabilize cooperation. If Adam knows that Eve will occasionally be stubborn enough to leave the refrigerator empty, he'll do the shopping more often himself as a precaution. The price, however, is that they argue on a regular basis. That's why the mixed strategy is only suited to situations in which little is at stake. As fatal as it proved for James Dean's rival, it is unthinkable that Adam and Eve would use it to decide who was going to pick up their child from day care.

Tit for Tat

Things get more complicated if a full refrigerator is desirable but also something our couple can do without. Then only one Nash equilibrium is available: Neither goes to the store, because no matter what one partner does, it's to the other's advantage to refuse to go.

And yet there must be ways to work together even in such a case, or our lives would be full of absurd situations. For example, whoever found himself to be the only customer in a bakery and was thinking only of his own immediate advantage would simply take his croissants and leave without paying. The baker could never prove that the customer didn't pay. But since the baker would know about this temptation, he wouldn't even wait on the solitary customer in the first place. That one of them wouldn't get any croissants and the other wouldn't get any money is in fact the Nash equilibrium in this situation.

But this description is unrealistic. The baker, after all, wants to keep his customer, and the customer would like to get his croissants the next time he visits the bakery. And as long as Adam and Eve stay together, there is hope that they will regulate their arguments about who does the shopping. The solution consists in both rewarding and punishing the other. If Adam goes shopping today, Eve will be willing to do it tomorrow. Then both can enjoy good food without having to put themselves out too much. But if he refuses to go, then she won't do it for him tomorrow.

Thus even complete egocentrics are capable of cooperation as long as retaliation is an option. A systematic study by the American political scientist Robert Axelrod came to the same conclusion in the early 1980s. To be sure, he didn't use real, especially ruthless human subjects, but simulated their behavior in a computer program. The virtual egocentrics were pitted against each other just as various chess-playing programs have played each other. Axelrod, however, used the example of Adam and Eve struggling over who would do the shopping. For historical reasons, the scenario is also known as the "prisoner's dilemma." In each round of the game, the virtual players had to decide for cooperation or non-cooperation. If both cooperated, each received three points. If both refused to cooperate, each received one point. The player that scored highest was the one who refused while the other trustingly chose cooperation. In that case, the sly player received five points and the trusting one got nothing. The opponents would play a couple of rounds against each other and then each would face a new opponent. In the end, the virtual player with the most points won.

What is the best strategy for this game? To always opt for

cooperation is impossible, because the good-natured player will be ruthlessly exploited. So it seems more promising to always decide to refuse cooperation. But the player who chooses that option only wins a lot of points until the others see through his game and refuse cooperation in return. The player who does best works together with partners willing to cooperate, but doesn't let himself be exploited by a sly player.

Axelrod invited experts to submit promising strategies and then staged a large-scale contest. Sixty-three programs were entered, many highly sophisticated. But to the surprise of most participants, the program that won was relatively simple. It was called Tit for Tat. Its strategy was to begin with cooperation and then simply copy the decisions of its opponent. If the opponent also decided to cooperate, Tit for Tat continues to do the same and both are rewarded. But if the opponent is treacherous and refuses cooperation, Tit for Tat responds in kind immediately.

Nevertheless, it is a gentle strategy. It survives in a world populated by complete egocentrics and yet shows characteristics reminiscent of selflessness. It is generous, optimistic, and forgiving. It's optimistic to attribute goodwill to your opponent until the opposite proves to be the case. That's also something we hardly expect from an egocentric. But Tit for Tat never begins with confrontation. The strategy only reacts with hostility once the opponent has evinced it. But as soon as the other returns to cooperation, Tit for Tat joins in. It behaves leniently, immediately forgiving the missteps of the other.

The success of generosity is perhaps the most astonishing result of Axelrod's simulation. Tit for Tat can never score better than its opponent, since all it does is copy the opponent's moves. The royal road to success seems to be to never want more than the other guy and never, ever try to outsmart him.

Thus Axelrod reached the conclusion that wise egocentrics should be generous, benevolent, and forgiving.[7] There was a huge response to his results, since they seemed to answer the ancient question of where morality comes from. Apparently, pure individual self-interest produces good mores. The way seemed to be open for channeling egocentrism in such a manner that it not only brings maximum benefit for oneself but also for others. Many of Axelrod's contemporaries found that comforting. If we humans really are such selfish creatures, maybe that wasn't so bad after all.

A Brief Golden Age

Of course, further experiments showed that the Biblical "eye for an eye and tooth for a tooth" is by no means an unbeatable strategy. A strategy that overlooks an opponent's first attempt at cheating and waits until the second to strike back proved to be even better.

Life, after all, is full of misunderstandings. However, as soon as a strict tit-for-tat strategy misinterprets an innocent move as treacherous, it responds in kind. And if the opposing player follows the same strategy, he's also going to stop cooperating, and both players will get stuck in a long, destructive conflict. Shakespeare's Othello, too, ought to have given Desdemona a second chance before accusing her of adultery, strangling her in a jealous rage, and then killing himself in remorse. The more patient Tit for Tat lowers the frequency of such errors; getting exploited somewhat more often by one's opponent seems a reasonable price to pay for this advantage.

There was no such amiable program entered in Axelrod's first tournaments because no one ever imagined that it could be successful, which clearly shows how mistaken even expert intuition can sometimes be. Like all of us, even the professionals allowed themselves to be seduced by the prospect of quick reward and put their money on grabbing what they could right away.

Yet with this very human greed for quick profit, we block our own path to lasting success. That is the clear lesson of Axelrod's experiments.[8] What would happen, then, if word got around of the success of generosity and good-naturedness and everyone started to follow that winning strategy? Then no one would take advantage of another, since Tit for Tat never starts out with an unfair move. Yet sadly, this state of affairs can never be permanent. As Bertolt Brecht wrote ironically in his play *The Caucasian Chalk Circle*, "a brief Golden Age of—almost—justice" is about to begin. Here is the problem: If every-one behaves fairly and honestly, there is no more need to impose sanctions on malevolent behavior. It follows that the ability to iden-tify and punish freeloaders will disappear. It won't be needed, after all, and whatever is of no evolutionary advantage gets lost sooner or later. But in such a world, populated by naive and good-hearted crea-tures, the dishonest have an easy time of it. The gentle beings get robbed blind, the ruthless thrive and multiply, until someone finally starts playing Tit for Tat again and punishes the scoundrels.[9] And the cycle begins again. Perhaps this cycle is the reason that two separate communities so often oscillate between phases of mutual goodwill and mistrust.

Fine Antennae for Cheating

We live in a delicate balance between forbearance and greed, coopera-tion and ruthless striving for our own advantage. Moreover, this bal-ance is constantly shifting. That's why there is no hope that one fine day love alone will rule the world. But it is just as mistaken to be too pes-simistic about human nature. When the gentle of the earth begin to predominate, the unscrupulous see their chance. But on the other hand, even ruthlessness can be overcome with amiability.

And so, we have well-developed antennae for both the good and bad intentions of our fellows. Our brains are programmed to constantly question the trustworthiness of others. The psychologists Leda Cos-mides (Santa Barbara) and Gerd Gigerenzer (Berlin) conducted a strik-ing experiment on this theme,[10] one you can repeat yourself. Imagine that there are four slips of paper on a table in front of you. Each one has a numeral on one side and a letter on the other. The sides facing up show D, 7, 3, and F. Your task is to find out if this statement is true: "If a slip has a D on one side, it will have a 3 on the other side." You should test the hypothesis by turning over as few slips as possible. Which will you choose?

Our Antennae for Cheating: The four cards at the top of the page have a number on one side and a letter on the other. Using the smallest number of cards possible, which would you turn over to confirm the truth of the following rule? "If a card has a D on one side, it will have a 3 on the other."

Most people find it much easier to solve this puzzle if possible cheating is involved. In the example at the bottom of the page, you are to make sure that everyone who applies for a pension has worked at least ten years. Which folders will you open? The brain is obviously less well programmed for logic than for protection against being taken advantage of.

Only a fourth of the Stanford University students Cosmides and Gigerenzer used as subjects found the correct solution: D and 7. (If the 7 had a D on the other side, it would disprove the hypothesis.) But the same problem becomes quite easy if the situation is made less abstract: Imagine that you are a case worker for a pension fund. You have the following information about four members of the fund: "has worked eight years," "receives a pension," "does not receive a pension," "has worked twelve years." The rule is: "A member must have worked at least ten years to receive a pension." Now it's immediately clear that you only need to examine the first and the second cases to make sure no one is unfairly receiving a pension. For only the member who has worked a mere eight years and the member receiving a pension can possibly have violated the rule and cheated the system.

Obviously, our minds have more or less difficulty solving a problem depending on how it is packaged. Cosmides and Gigerenzer concluded that the human brain is not particularly well suited to abstract, logical thought but is much better at detecting fraud. A little control experiment confirmed their results. When the psychologists asked their subjects to approach the same problem but imagine themselves as retirees instead of case workers, they again responded promptly but in a completely different way. This time, almost all the subjects recommended taking a closer look at "does not receive a pension" and "has worked twelve years" since what is now at issue is whether the fund has unfairly denied someone a pension. It didn't occur to anyone that these two cases have no bearing on the validity of the rule ("A member must have worked at least ten years to receive a pension"). If we're looking for fraud, it doesn't seem to matter that logically, the man without a pension and the woman who has worked for twelve years perhaps simply forgot to apply for their benefit.

One Hand Washes the Other

According to game theory we are cold-blooded calculators—or should be if we want the greatest possible advantage for ourselves. John von

Neumann, for whom it seemed inarguably logical to recommend a first strike with nuclear weapons, could hardly have imagined where his ideas would lead once we continue to follow their logic. For it is none other than his theory that constitutes the foundation for an insight that would have seemed absurd to him: Tolerance and generosity are in many situations the only reasonable choices.

The influential American evolutionary biologist Robert Trivers dubbed this strategy "reciprocal altruism,"[11] and the term has become established, although it's misleading. For in reality, players are not acting selflessly; they expect that the other will reward them for their indulgence. The epigraph of this chapter describes this attitude by reversing the usual associations. He who gives does so from gratitude; he who takes recognizes and honors the other's act.

Our reflexive vigilance against attempts to cheat us suggests that humans not only follow this strategy consciously, but have also internalized it. We automatically prick up our ears at the slightest hint that someone could be taking advantage of our trust, just as we automatically wince at an unexpectedly loud noise.

Betrayal is more likely the less stable a relationship is. For reciprocal altruism basically speculates on a trade-off: I'll let you go first today so that I can benefit from you tomorrow. Reciprocal altruism can only work if there is a great likelihood that one's own concession will pay off.

By contrast, no one would expect much from a casual vacation acquaintance. We are highly unlikely to forego something in favor of someone we will probably never see again. So the logic of mutual taking and giving has its disquieting side. For example, what happens in a relationship when the partners think of each other as what Germans have ironically dubbed *Lebensabschnittsbegleiter* (companions for one phase of their lives)? In any event, there won't be much appeal in the idea of occasionally putting your partner's interests ahead of your own. According to the principle of reciprocal altruism, the mere thought that he or she might move out in the foreseeable future keeps us from doing too much for them, which of course makes the relationship all the more unsatisfactory and precarious. "As soon as we become convinced love is not possible, love becomes impossible," as the American evolutionary biologist and psychiatrist Randolph M. Nesse once put it.[12] Conversely, the wedding vows that sound so old-fashioned to many could in reality be quite rational. By pledging themselves to be together "till death do us part," newlyweds lay a foundation for doing things for each other and thereby ensure a stable relationship.

It's the same story in the world of work. Employees who fear they will be expendable in the next reduction in force would be acting irrationally if they did more for the company than what's expected of them, and thus they make themselves expendable in fact. Although flexibility in hiring and firing may benefit a company in the short run, in the long run it's highly probable that both sides will be worse off.

It's equally risky for a society to allow the gap between its poorer and richer members to become too great. For even if the poorer ones are not objectively suffering any deprivations, too great a disparity puts a great strain on *everyone's* willingness to cooperate and be generous and forgiving. According to the principle of reciprocal altruism, people are more willing to cooperate the more they have to exchange. But what can people give one another if some have nothing to give and others can already afford anything they want?

CHAPTER 3

Building Trust

He who does not trust enough will not be trusted.

LAO-TSE, *TAO TE CHING*

AT FIRST IT WAS ONLY THE WOMEN, returning home with unusually light baskets. No one recognized it as the first sign of a change that would soon engulf them all. Instead, the Ik attributed the fact that the women were returning from the bush with fewer roots and berries than usual to the drought. Those were hard times for the East African people. Not only was there no rain for their fields, but they had also lost their hunting grounds when the Ugandan government turned the fertile Kidepo Valley and the surrounding mountains into a national park.

Now when the men went hunting, they had to do it illegally. They hunted for days in search of a gnu or a gazelle, for their prey was starving as well. Whatever the hunters killed would be divided among them. At least, that was the custom among the Ik. But lately, individual hunters seemed to disappear into the bush and not respond to even the loudest shouts. It was as if the mountains had swallowed them up. But when the hunting party returned to the village, they found the defectors alive and well in their houses. They avoided saying where they had been.

When more and more men started disappearing from hunting parties, the Ik began to understand. The absentees had been able to kill a prey on their own and then conceal the animal in order to keep it all for themselves. If there were any leftovers, some even attempted to sell them. And now the men also understood that their wives were doing the same thing. The baskets weren't empty because there were no more roots and berries, but because the foragers were no longer sharing what they found with the others.

This is the tale told by the British anthropologist Colin M. Turnbull, who lived among the Ik. He is the sole witness to the catastrophe that befell the Ik in the years after 1965. At that time, hardly any foreigners ventured into the trackless mountains between Uganda and Kenya, where armed cattle thieves roamed and tribal feuds raged. Turnbull was not just some run-of-the-mill anthropologist. He had made his reputation at Oxford, written a well-received study on Pygmies in the Congo, and risen to the post of curator for African ethnology at the New York Museum of Natural History. The opportunity to compare the almost unknown Ik hunter-gatherers of East Africa with the Pygmies he knew so well came about unexpectedly, after plans for two other field projects fell through. In the Kenyan port of Mombasa, Turnbull bought a fire-engine-red Land Rover—the only one available—and set off. He did not know that he was on his way to a society in the process of disintegration.

Not that they were bad people. The Ik had always been peace-loving and remained so even in the worst years of famine. Nor did Turnbull ever observe any signs of hatred. Rather, the people simply ceased having any interest in one another. Stillness settled on the villages as the Ik lost the habit of conversation. They said only the minimum necessary to avoid misunderstandings. Whoever had nothing to do would join the others at the highest point in the village, sit down, and stare, brooding, into the distance.

The People without Trust

Turnbull drew a controversial conclusion from his observations: The Ik had shown each other the true face of humanity. The fate of this people proved that sympathy and goodness are not part of our makeup but rather a masquerade we keep up only as long as we can afford to—or as long as it's easy to do so.

When Turnbull published his report in 1972, it created an immediate sensation, although other researchers soon criticized his arguments and cast doubt on his observations.[1] *The Mountain People* was a worldwide bestseller and won prestigious awards. The English director Peter Brook turned it into a play entitled *Ik* that was seen by enthusiastic audiences in Paris, London, Vienna, and New York. Readers and theater-goers, by turns fascinated and horrified, obviously thought they were seeing their own fellow humans in the Ik. And that was Turnbull's intention: He wrote that the Ik "had cultivated individualism to its apex."[2]

Does Turnbull's study in fact reveal our true nature? Isn't it the report of a metamorphosis instead? He depicts a society that disintegrated because its people gradually lost trust in one another. When times became tougher, people presumably were tempted to think of themselves first, and gradually more and more followed suit. The men who devoured their prey while out hunting, the women who stuffed themselves with berries while still picking—none of them believed any longer that it was worthwhile to share with others. At the moment, each of them was behaving quite rationally. The Ik acted exactly as the game theory described in the previous chapter would recommend. After all, the principle of reciprocal altruism suggests that if one is getting little, one should give little in return. The problem is that others respond with the same behavior, and that sets off a downward spiral. This was the real tragedy of the Ik: In a crisis in which people had greater need of one another than ever, they abandoned one another, which only accelerated their doom.

The Ik and the Bankers

As strange as the Ik seem to us, they must seem quite familiar to any economist. Probably unconsciously, Turnbull describes them as living embodiments of those abstractions that populate economics textbooks. *Homo economicus* has been much maligned. He is a conjectural human who behaves with complete rationality every second of his existence, ready to do anything for a reward but nothing without one. You don't want to meet up with him, not even in your worst nightmare, because *Homo economicus*, as the London economic journalist Tim Harford has written, "would strangle his own grandmother for a dollar—assuming it didn't take more than a dollar's worth of time, of course."[3] Turnbull's Ik come very close to looking like this unpleasant fellow. In fact, a later chapter of his book depicts how the Ik abandon the old and the sick to their fate and how an old man starves to death because his children snatch the food from his plate.

Of course, economists don't really believe that *Homo economicus* corresponds to humanity's true nature. They usually explain that he is simply hypothetical, a caricature they need in order to conduct certain analyses. But the trouble is, so far the economists have no more accurate image of human nature with which to replace their straw man. So when they advise politicians, make recommendations to managers, or appear on talk shows, their view of things is based on the assumption that people really do behave like the Ik.

And from time to time, their scenario becomes reality. In the days after the September 2008 collapse of Lehman Brothers in New York, when other banks' solvency seemed uncertain, suddenly no bank would loan any money. The world economy teetered on the brink. And it is certain that within a few days, one bank after another would have collapsed if the governments of the developed countries had not pumped several trillion dollars into the markets. The banks themselves, forced to follow the rules of reciprocal altruism, were incapable of saving themselves. Whoever follows game theory must always strive for maximum advantage for herself—even if she knows very well that she too will lose if everyone does the same.

It later emerged that the crisis was caused by the banks' thoroughly abusing the trust of their customers by selling trillions of dollars in bad loans. Motivated by the prospect of generous bonuses, many investment bankers acted precisely like *Homo economicus*.

Reciprocal altruism is necessary to protect the trusting from exploitation. However, this exchange of favors doesn't appear to be very reliable if it breaks down in times of crisis—just when you're depending on it the most. Thus the Ik headed toward a catastrophic famine. The horrifying scenes Turnbull describes of relatives stealing food from one another must have been due to people's desperation. Whoever has to fight for every mouthful forgets everything else. In any event, the Ik were incapable of freeing themselves from their desperate situation; only aid from outside allowed them to survive. The rescue mission was successful. In a dozen villages today, you can hear people speaking Ice-tot, the language of the Ik.

Feelings of Solidarity

Thus there are limits to the effectiveness of reciprocal altruism. Obviously, this principle is not enough to explain cooperation among humans and their willingness to help one another. If people really always looked only to their own advantage, our ancestors would hardly have been able to survive hard times. Societies whose cohesiveness beyond the bonds of family depends on the exchange of favors have little chance of success in times of crisis. In any case, successful models in nature look different. So the question is whether there is something else that guides our life together besides the care for our relatives and reciprocal altruism.

Such a principle—beyond what traditional game theory has to offer—is trust. With it, humans progress beyond the purely rational rule "I'll treat you the way you treat me." They give someone else something and hope that he will later reciprocate. But they can neither force the other to return the favor nor punish him if he doesn't. Trusting means exposing yourself to risk.

And yet not a day goes by when we don't chance it. A student buys a used computer from a stranger on eBay, paying what for him is a significant amount of money, and can only hope that the stranger delivers a functioning machine. A company pays for an employee's advanced training because it believes she will reciprocate with increased effort. However, if the employee moves to a competing firm shortly thereafter, it's tough luck for the former employer. And the student won't see his money again if the computer gives up the ghost two weeks after he bought it, since in a private transaction the seller cannot be held responsible for the quality of what he has sold. Nevertheless, far more than 10 million articles are sold on eBay every month in Germany alone—an astonishing statistic.

Even in the supposedly cynical atmosphere of the prisoner's dilemma, people trust one another. The previous chapter has described this strategy using the example of Adam and Eve's conflict about grocery shopping. In the original version that gave the experiment its name, two partners in crime have been arrested and are being held in separate cells. They've robbed a bank but know that the only thing the police can prove is illegal possession of firearms. During interrogation, the police offer each one a deal: Whoever testifies against his partner will get a suspended sentence while the other will get fifteen years. If both turn state's evidence, however, they both have to serve ten years. The other five will be suspended for their cooperation with the authorities. But if both of them refuse to cooperate, they can only be convicted of the lesser charge of firearms possession and will serve only a year behind bars.

Would you cooperate with the authorities? Imagine you are the second to be questioned. They tell you what your partner has decided to do. If he sang and you do the same, at least you'll save yourself five years in jail. But if your buddy was a stand-up guy and kept his mouth shut, then you can escape any stay in jail. All you have to do is betray him.

But if you're the first one to be interrogated, it's obvious that there's nothing in it for your partner to stand by you. After all, you'll probably never see each other again. It follows logically that the only rational choice for you if you want to spend as little time as possible behind bars is to talk.

If human subjects are asked to put themselves into this hypothetical situation and figure out how to spend the fewest years in jail, they're capable of determining the proper decision. More than 80 percent of participants choose to snitch, showing that they understand the logic of the game very well.[4]

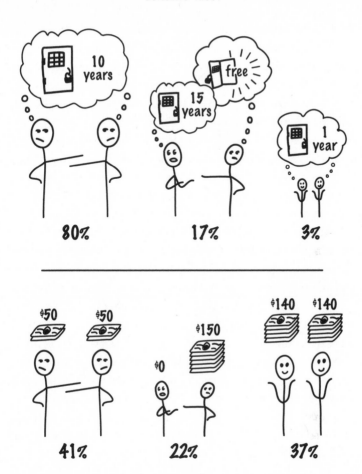

The prisoner's dilemma. Above: Two partners in crime are interrogated in succession. They have robbed a bank but know that the authorities can only prove that they are in illegal possession of firearms. If both testify against each other, each will get ten years. If the first one refuses to cooperate but the second one testifies, the first gets fifteen years and the second gets a suspended sentence. If both refuse to cooperate, each only serves one year. When asked what they would do in this hypothetical situation, 80 percent of people say they would choose to betray their partner. For egocentrics, this is the only logical answer, although it is the worst outcome for both partners.

Below: If subjects are offered money ($10 for every year they avoid jail), the number of traitors is almost halved. Now 37 percent of those asked would maintain solidarity with their partner. The experiment shows how much our willingness to act selflessly depends on the situation. In the purely intellectual prisoner's dilemma game, people choose according to pure reason, but when money is involved, they have recourse to their experience that it is worthwhile to be trusting.

But what do subjects do when, in the game, they get $10 for every year they don't have to spend in jail? You might think they would be even more motivated to turn in their partner. But the opposite is true. The number of stand-up guys rises dramatically. To be precise: 58 percent of the first to be interrogated are prepared to bet on the solidarity of their partner. The second to be questioned were again informed what the first ones had chosen. If the first were loyal to the second, 62 percent of the latter rewarded that behavior by voluntarily foregoing the extra money, although they could have betrayed the first without any risk and collected $150 instead of just $100. Thus a new balance is achieved that we can't explain with the logical rules of game theory.

Asked the reason for their illogical behavior, the second players answered that they simply liked it better that way. It was more satisfying to cooperate with the first players than to betray them—even if trust is risky and gratitude cost them money. Personal sympathy played no role, since participants didn't know each other, never met, and communicated only via computer. Obviously, something needs to be at stake to make solidarity seem worthwhile. As long as winning meant nothing more than getting an abstract point total, participants made the purely logical choice. When a concrete reward was involved, their feelings won out.

Cooperation Makes for Happiness

Scientists can show us a picture of what is going on in the heads of such trusting individuals. For the last twenty years or so, magnetic resonance imaging has allowed them to watch the brain as it thinks and feels. Subjects have their heads placed in a tube like the ones many hospitals use in computer tomography. While a person solves a given task, the device measures which parts of the brain have an increased blood supply, indicating heightened activity.

The American anthropologist James Rilling used this technique to observe the brains of thirty-six women while they were working on the prisoner's dilemma.[5] Whenever both participants in the game trusted each other and opted for solidarity, brain centers were activated that generated good feelings. Cooperation made them happy—the participants didn't just say so, they really felt it.

The brain system responsible for that is evolutionarily one of the most ancient. It consists of a widely branching web of gray cells that originates in the mid-brain and whose main function is to lure us into promising situations. Whether it's a box of chocolates smiling at us from a display case, our boss unexpectedly promising us a raise, or an attractive potential sexual partner walking by—the so-called reward system springs into action. At first it steers our attention toward promising things or people. Then it sees to it that we feel pleasure or a pleasant anticipation. And finally, it helps us to learn from our experiences. Unexpectedly pleasant experiences activate our memory, for we are meant to imprint thereon what or who it was that made us feel good so that we can produce the same situation the next time we have the chance.[6]

That's why trust can produce euphoria; someone else did more for us than we expected.[7] The reward system then causes us to remember what we have registered about our well-meaning fellow human. Whoever acts fairly leaves a deep trace in our memory. Although it's true that at the movies, the bad guys are often the most interesting characters, apparently they only occupy our attention for a brief time. Experiments have shown that we remember people with whom we have had a good experience much better than people who have cheated us.[8] Brain research is also able to explain why players in the prisoner's dilemma trusted each other only when money became a factor: The reward system—which doesn't follow logic but rather our earlier, emotionally determined experiences—influences our behavior more strongly if there's a possibility of a reward. Otherwise, cold calculation predominates.

Thus our preferences are more flexible than the theories of the economists admit. Before every decision, *Homo economicus* supposedly calculates precisely what profit each option will produce, and then chooses the most profitable. But flesh-and-blood humans are almost never so thorough, because they cannot be. Life is full of imponderables, and we usually have to decide quickly. That's why we prefer experience to logic and orient ourselves according to what worked well in the past, even if it's not the most logical in a given case. Trust is simply born of this pragmatic attitude.[9]

When the anthropologist Rilling explained to his subjects that their opponent was really a computer, the women were no longer ready to trust—although the computer was programmed so that it would react just like a human being. But who could blame the participants for their mistrust? No one has ever heard of a thankful computer. And finally, a further variation in the experiment proved that a pleasurable experience wasn't just a matter of money for the players. When Rilling let them win the same amount in a game of solitaire as they had in the prisoner's dilemma, the reward system in their heads reacted much more weakly.

All these results make a mockery of *Homo economicus*. According to traditional economic assumptions, the only thing that should interest us is our concrete advantage, no matter how it is achieved. But in reality, a dollar is not always a dollar. Much more important than what you can buy with the dollar is the story you attach to it.[10] Not least of all, the experiments contradict the fanatical disciples of competition, for they provide no indication that humans find their fulfillment in triumphing over others. It's not competition for its own sake that we seek, but rather cooperation.[11] What we achieve together with others obviously makes us happier than what we achieve against others or by ourselves.

Brains in Sync

Thus humans are trusting in principle. Yet we would hardly be willing to count unconditionally on every random contemporary we meet, especially if we're risking something more than a few dollars in a scientific experiment on game theory.

Numerous people have tried to explain why we are willing to entrust our child to one person but wouldn't trust another to give us the time of day. George W. Bush, for example, had one of the more odd opinions about what trust is based on. When the former president met Vladimir Putin for the first time, Bush said he had looked the Russian in the eye and "was able to get a sense of his soul." That convinced him that Putin (a man who had made his career in the KGB) was "straightforward and trustworthy."[12] At their next meeting, held this time at Bush's ranch in Texas, the host didn't think it necessary to get an agreement on the mutual destruction of five thousand atomic warheads down on paper. Putin however rightly insisted that there had to be a signed agreement and controls. Soon thereafter, Bush invaded Iraq, and at the next summit conference he was no longer so sure of Putin's cordial friendship, since the Russian leader opposed the march on Baghdad.

In order to achieve more convincing assertions about how this special relationship of trust develops between two people, Kevin McCabe invented the "trust game."[13] The American McCabe, Professor of Economics and Law at George Mason University, is one of a growing number of economists who are dissatisfied with the unrealistic theories of their colleagues and want to discover through experiments how people really behave. The trust game is easy to explain. The first player (the giver) gets a sum of money and can freely decide what share of it he wants to give to the second player (the receiver). The investigator then deposits twice that amount in the receiver's account. The receiver can now decide how much of that profit she wants to return to the giver. Thus the first player in effect extends credit to the second, but without any claim to reimbursement. In real life, the giver is like a student who purchases a used computer on eBay, while the receiver is like the seller.

Basically, the trust game functions like the prisoner's dilemma. If the partners work together, both come out ahead. If the second player is stingy, she ends up keeping more and the first suffers a corresponding loss. If both refuse to relinquish anything, no one gets anything. But in the prisoner's dilemma the players have only two choices: keep silent or betray the other. The trust game allows for much finer gradations of how much they risk and what they're willing to do for each other.

Of course here too, the logic of game theory dictates total noncooperation. The giver would not be allowed to part with a cent, since he could not expect to get anything in return from the receiver. Yet again, humans do not heed the advice of their reason. When McCabe supplied his givers with ten dollars, the subjects entrusted on average five dollars and sixteen cents to their unknown partners. Some even gave the entire ten dollars. Their expectations were mostly disappointed, however. Whereas in the prisoner's dilemma, hardly anyone dares to betray a partner who has trusted him, the trust game is designed to encourage stinginess. Although generously endowed receivers were willing to acknowledge their benefactors, they did so in a mingy way. On average, they returned four dollars and sixty-six cents—more than they had to, but less than they had received.

On the other hand, the receivers were much more generous once both partners had played several rounds with each other. For now they had to weigh their desire to keep as much as possible for themselves against the need to return enough to the giver to keep him in a giving mood. After all, they wanted the giver to part with the largest amount possible in the next round. The task was to figure out how little was still acceptable to the giver. And how much does one have to give back to tempt the giver to give even more?

The receivers couldn't afford to annoy their partners. And so they would return unusually large amounts the moment a giver suddenly began to reduce her payments; they took the reductions as a sign that she was dissatisfied. And usually, this placating strategy worked: The giver again believed in the trustworthiness of the receiver and raised the payments again. The partners were playing a very sophisticated form of tit-for-tat. In the end, they succeeded in maintaining a stable balance of exchange over the entire game.[14] Moreover, much more money was exchanged—to the advantage of both players—than was the case when players interacted only once and then never again.

Since the players are constantly testing each other, their relationship follows a course as uncertain as a chess game. Brooks King-Casas, a young behavioral scientist in Houston, has researched what goes on in the heads of such players in a new kind of experiment.[15] He paired up players of the trust game on the Internet. One partner was in Texas and the other in Pasadena, and while they played, King-Casas and his team monitored their brain activity with an MRI scanner.

The subjects knew nothing about their partners except how much money they were sending. They could read the amount on a screen in the scanner. What the researchers saw on their monitors were two brains in sync. Separated by fifteen hundred miles, the players' brains reflected not only each other's decisions but even their intentions. After a few rounds the subjects knew each other so well they could predict how much trust the other would extend the next time around.

In fact, the activity in the two brains soon looked as though the participants could read each other's minds. Once the givers made their decision known, the more money they entrusted to the other, the stronger the activity in certain centers in the so-called *Gyrus cinguli* deep in the cerebrum. At first the receiver took a few seconds to react to the giver's decision. Then another core in the receiver's cerebrum called the *Nucleus caudatus*, part of the reward system, would go into action. The stronger its signal, the more dollars the receiver would send back. And since the latter almost always rewarded the giver for her trust, the two brains were operating in synchrony: The giver's brain made its decision and ten seconds later the receiver's brain did the same.

After a few rounds, however, this sequence was reversed: The decision signal in the receiver's brain now occurred four seconds *before* he learned how much the giver would send him. The receiver's *Nucleus caudatus* had already decided whether the giver would be generous and deserve a large reward. And the receiver's brain was almost always right. In a follow-up experiment, King-Casas asked receivers to estimate the dollar amount of the payment they expected. Their estimates were often exact to the last dollar, and they improved steadily the longer the game went on. Thus to trust others is first of all to know them well.

The players based their strategy on this mostly unconscious knowledge. Whenever the partner behaved as expected, they would pay him as much in the next round as they had in the previous one. If he exceeded their expectations, he got more the next time. But if he disappointed their expectations, he was kept on short rations. Thus if all goes well, a spiral of trust is established. Round for round, the partners can become more and more generous with each other.

In Praise of Blind Trust

If someone displays repeated generosity, less and less caution seems required, until at last, a relationship with them enters an entirely new phase that is astonishing from the point of view of game theory: The partners trust each other blindly.

Of course, not everyone is capable of total, unreserved trust. When Kevin McCabe had twenty-two pairs of players repeat the trust game several times with each other, only half of them achieved that relaxed state, even though they were permitted to establish a certain personal relationship. Although they did not meet each other directly, they were shown a photograph of their partner.[16] The eleven pairs that did reach the highest level of trust were at first distinguished by special vigilance. In these players, the brain's reward system, which compares the expected with the actual payments, displayed heightened activity during the initial rounds. Naturally, at the beginning of the game these people were trying very hard to test the credibility of their partners and imagine what they were like. And so these subjects consistently resisted the temptation to sometimes pay their partner nothing. At the end of the game, the accounts of both players were full of money.

Once these players seemed to know their partners well enough after a few rounds, they leaned back, as it were. The activity of the reward system slowed down, and in general, the trust they had established saved them much time and energy. As the experiment demonstrated, these players made their decisions much more quickly and with less brain activity than other subjects who remained distrustful. Control is good, but trust is better.

On the other hand, in these relaxed players activity increased in regions concealed in the center of the cerebrum, where our more intimate social ties are regulated. For example, these areas of the brain give a signal when people look at their babies or their loved ones.[17] And again, the brains of both partners were operating in sync. But while the systems that attempted to predict the other's next move had become synchronized as the partners were getting acquainted, now the most activity was in centers that regulate sympathy and emotional intimacy. It seemed that the partner's self-interest was no longer a source of doubt.

Later, the trusting participants reported that in fact, they had felt friendly toward each other from a certain point on, even though the only things they knew about each other were how they looked and how much money they were willing to give. Such feelings owe their existence to the *Regio septalis*, where there was increased activity in the second phase of the game and the players' brains were again acting in sync. There the neurohormones oxytocin and vasopressin work to make us more affectionate toward one another. These neurotransmitters are always active when sympathy motivates us to care for others. Without them we would neither enter into romantic relationships nor love our children. Trust in others is thus a product of elemental functions of the brain to which we owe our very survival.

Therefore, a functioning, mutually beneficial relationship is by no means merely the result of rational choice; it rests on an emotional foundation. That was also confirmed in a study conducted by the Zurich neuroeconomist Michael Kosfeld. He administered oxytocin to male subjects in a nasal spray to activate the corresponding regions of the brain. Suddenly his subjects' belief in the goodness of others seemed to increase, and they were prepared to give away much larger amounts in the trust game.[18] Disturbingly enough, entrepreneurs are already attempting to cash in on such experimental results, which we will look at more closely in Chapter 5. A U.S. company with the dubious name Vero Labs is advertising a product on the Internet called Liquid Trust. It comes in small but expensive dispensers that look something like deodorant, and it supposedly contains oxytocin. You spray yourself with it after your morning shower.

"Trust is power," as the website quotes the Chinese philosopher Lao-Tse. This piece of ancient wisdom will supposedly help singles who, thanks to Liquid Trust, will no longer need a high-powered job or car to inspire trust. Vero Labs claims that users will gain the trust of women and find they have better luck with them, and that savvy salesmen who dab a little Liquid Trust on their correspondence will do better business.[19] To be sure, it doesn't inspire much trust that the website neglects to inform us that the concentration of hormone in their sprayers is orders of magnitude weaker than as a nasal spray. Nor do they say anything at all about the fact that even when oxytocin is administered nasally, its effect, while measurable, is by no means overwhelming. The truth is that people inspire trust in others by being trustworthy.

Homo Economicus Is a Bad Businessman

Half of the pairs of subjects in McCabe's experiment failed to gain each other's trust. Both partners remained suspicious of each other to the end. The reward systems in their heads even became more active over time, as if each were desperately trying to assess the other. But as the subjects themselves explained after the experiment, they never stopped perceiving the other as an egocentric person, an enemy. And thus they repeatedly experienced breaches of trust resulting in very meager account balances for both.

Whoever wants to build trust has to be able to read correctly the signals others send. An impatient look or a skeptical undertone can be such a signal, but so can the flow of payments, as in McCabe's experiment. If it peters out, either the partner is not trustworthy or the relationship is slipping into a crisis.[20] But people who miss such signals can never cash in on the returns from trust. That's why *Homo economicus* is a bad businessman: He completely ignores the emotional life of his fellow business partners.

But there's more: You can test the trustworthiness of others only if you yourself are able to trust. That is the conclusion reached by King-Casas in another fascinating study.[21] He recruited men and women who were overtly incapable of repairing damaged trust and often didn't notice when someone was beginning to doubt their good intentions. People who suffer from borderline personality disorder (BPD), for instance, have great difficulty in such situations. BPD patients suffer strong mood swings and have trouble building stable relationships.

This trouble also emerged when King-Casas asked them to play the trust game. If the receivers had BPD, the givers soon began to send less and less money even though ignorant of the receivers' disorder. The reason was that the BPD patients misunderstood signals of annoyance and neglected to placate their partners. Non-BPD sufferers would mollify their givers by temporarily raising their payments when the former began to withdraw from cooperation by giving less. The BPD patients, on the other hand, would give less in their turn and thus escalate the crisis. Cooperation would continue to erode.

The BPD patients engendered mistrust because they themselves didn't trust their partners. That was evident when King-Casas measured his subjects' brain activity. When a non-BPD receiver got an unexpectedly low payment, there was a reaction in a part of the cerebrum called the insular cortex, which normally signals physical pain. However, the insular cortex also sends out signals when we feel we have been badly treated. The brain processes psychological pain just as it does physical pain. This discomfort signaled to the non-BPD subjects that something was amiss in their relationship with their partner, who had violated the norms of their exchange. Such a breach of trust causes pain.

In the head of the BPD players, on the other hand, the insular cortex remained inactive when their partner suddenly became stingy. The BPD players were spared the painful but very useful signal because they expected no cooperation from the other. After all, they themselves were unusually distrustful, as became evident in their answers to a series of questions after the game was over. So why should their partners trust them? Since all these considerations remain unconscious, the distrustful players had little to show at the end of the game but never understood why.

Trust Brings Riches and Success

Trust brings more rewards to the trusting than even philanthropists would have believed. But in recent times, researchers have provided much unexpected evidence. For example, the American educational researcher Roger D. Goddard discovered that students who trust their fellow students do better on exams.[22] These young men and women cooperate better with their fellows, are more ready to share information, and thus learn more. Less trusting students can only count on themselves.

On a large scale, the prosperity of an entire country can be partially explained by how much the inhabitants trust one another. To test this hypothesis, citizens of thirty-seven countries were asked whether in their opinion one "can trust most people." Sixty-one percent of Norwegians agreed, but only five and a half percent of Peruvians did. Germany lay between those extremes with barely thirty-five percent.

As a rule, people were quite good at estimating their countrymen. In interviews, Norwegians disapproved much more frequently than Germans did of people obtaining social service fraudulently, hopping a ride on public transportation without paying, or keeping a wallet they've found. According to the American economists Stephen Knack and Philip Keefer, people aren't just paying lip service with these estimates.[23] For wherever people look the other way when something unfair is happening, they will also act unfairly themselves. This was the result reached by the researchers after they left wallets with money on the street and observed what happened.

It is not ruthless self-enrichment that makes a society prosperous. It is trust. National economies in which people believe one another to be generous consistently grow faster than others. Knack even provides a rule of thumb for measuring the connection: If a mere seven percent more citizens agree with the statement that one can trust most people, it translates into a one-percent increase in annual economic growth.[24] He says that this formula was valid at least for the period from 1970 to 1992. And by no means is it the case that people are only nice to one another when the factories are at full capacity and everyone's doing well economically. It's the other way around: It's easier to do business when people trust one another. Those are the conditions in which people invest more, and in the end, they enable more prosperity for everyone.

Thus even economic data make the ideas of traditional economic theory look implausible. Humans are certainly not beings who at every moment are striving for their own advantage. On the contrary, it is precisely because we are sometimes willing to put ourselves second in order to nourish a relationship that we are successful.

At the same time, our ability to trust one another is a dramatic refutation of the erroneous idea that altruism and egocentrism are irreconcilable polar opposites. In reality, each needs and is determined by the other. Trusting means acting selflessly, giving without expectation of a return. But such readiness to sacrifice can only survive if it pays off in the long run. That's the apparent paradox of trust: We often are pursuing our own interests most effectively by laying them aside and serving others.

CHAPTER 4

Feelings Without Borders

His surroundings had become so familiar to him that without noticing it, he was assuming some of the habits of the people who lived here.

GEORGES SIMENON,
MAIGRET AND THE MAID (FELICE EST LÀ)

VITTORIO GALLESE MADE THE DISCOVERY OF HIS life while working as a prison doctor. What he really wanted to do was research, but since he couldn't get a position at the university in his hometown of Parma, Italy, the young doctor put bread on the table by working in a prison. He was on service at night and on the weekends, treating hardened criminals. During the day he worked without compensation in his lab. Of course, he knew all about what his patients had done. The whole town knew their stories, because everyone convicted of a major crime anywhere in the province did their time in Parma. But Gallese felt no revulsion for the prisoners. On the contrary, he felt sympathy for them because they were ill, even if the patient happened to be a hit man or a serial killer who had dissolved his victims in acid. The guards were always asking him why he put himself out for these criminals, but he couldn't explain his feelings.

It wasn't until many years later that he was able to describe what must have been going on inside him at the time. "If I'd only read about the criminals in the news, I too probably would have felt nothing but repugnance for the killers. But those men stood before me in the flesh, talked about their wives, had a personal history as I did," Gallese told me when I visited him at the University of Parma in 2007. In the meantime he had become a professor. "They weren't completely alien beings. And, not least of all, we shared an environment. Seven doors closed behind me on the way from the street to my office; I knew what it was like to be cut off from the outside world. Because I ultimately lived with them, it wasn't hard for me to put myself into my patients' shoes. . . . As a doctor, I was there to heal, not to judge."[1] In his free time, Gallese did research with rhesus monkeys. To find out how the cerebrum gives instructions to the muscles, he and his colleagues had attached electrodes to some gray cells in the monkeys' brains. (The animals couldn't feel anything because the brain is not susceptible to pain. Still, any animal used for research suffers from being caged its whole life. The important insights that scientists have gained in this and other experiments involving animals in captivity come at a cost that requires serious ethical consideration.) Whenever the monkeys reached for food, certain brain cells were activated. The researchers' instruments would start crackling. But then something strange occurred. "When I myself at one point extended my arm toward the nuts," said Gallese, "the crackle occurred too—as if the monkey had moved. But it was only watching quietly. At first, of course, we thought there had been a mistake." But the signal was repeated every time the animal saw Gallese reaching for food. "After a while we realized that the monkey's brain actually behaves as if it were putting itself in our shoes. When an animal observes another's movements, the observer's neurons mirror the other's behavior. That's why we called them mirror neurons."[2] It took Gallese years to understand the connection between the signals from the brains of his monkeys and his feelings as a prison doctor. For a long time, he and his colleagues thought only that they had discovered an astonishing specialization of the parts of the brain that regulate movement. The monkeys clearly owed their gift for imitation to the mirror neurons. These cells were probably also responsible for an amusing peculiarity of newborn rhesus monkeys: If you make a face or stick out your tongue in front of the tiny animals, barely bigger than a human hand, they will imitate you.

Later it turned out that the mirror neurons can do even more. They also helped the monkeys recognize *why* another being is making a particular movement. Some of the neurons fire as soon as a monkey guesses that a human is reaching for nuts, even if the animal cannot see the food. Different mirror neurons turn on when the human is reaching for a bottle instead of nuts, although the gesture is the same.[3] The animal guesses the human's intention as though reading his mind.

Was the same mechanism at work in human brains? Everything suggested it was. Yet it was not until the spring of 2010, fifteen years after Gallese's first publications on the topic, that neurophysiologists in California succeeded in identifying individual mirror neurons in the human brain.[4] What is more, however, they discovered the curious gray cells not only in the centers that control movement, but also in an unexpected location: in parts of the cerebral cortex responsible for memory.

This seems convincing proof that mirror neurons are involved in much more than just our motor functions. Rather, these cells also mirror others' feelings. When we see—or even just hear about—someone else suffering pain, our brains react as though our own body were feeling it, too.

The neurons that Gallese and his colleagues first discovered in rhesus monkeys are much more than just a mirror. With their help, we experience the suffering of others as our own, as though the border between self and other were temporarily blurred. And since this kind of empathy is involuntary, it makes no difference if the other is a friendly neighbor or one of Gallese's killers.

Empathy Creates Trust

Leading brain researchers ranked the discovery of mirror neurons alongside that of the genetic function of DNA.[5] And it really does upend many traditional notions of how our life together functions. Sympathy, for example, is not produced in order to make others obligated to us by doing them a good deed; it happens involuntarily. That does away with the suspicion that neo-Darwinists like the American author Robert

Wright have promulgated as a piece of cynical wisdom: "The more desperate the plight of the beneficiary, the larger the I.O.U. Exquisitely sensitive sympathy is just highly nuanced investment advice."[6] The neo-Darwinists were convinced that a truly sympathetic person would always be at a disadvantage to someone who was hard-hearted and calculating. But isn't it possible that the opposite is true? As we have seen in the preceding chapter, success often depends on the ability to engender trust, and only people who understand their interlocutors can gain their trust. Nor is it enough to understand intellectually what motivates the other. Rather, our frequently invoked "social intelligence" is based on a talent for putting ourselves in the other's shoes emotionally. The trust game we examined reveals how resonance can develop between the brains of two players and help both to succeed. The mirror neurons play a major role in this phenomenon.

These gray cells themselves are only part of a recently discovered empathic system. Its circuits operate in a quite different way than strategic intelligence. They ensure that we are "infected" by others' emotions, that we can imagine ourselves in their place, that we understand our fellow humans and feel sympathy for them. All of these various impulses constitute empathy—our ability to put ourselves inside others' circumstances. Without empathy, our affections, communal life, and willingness to help one another would be unthinkable. Cooperation and trust both depend upon it.

Contrary to what people have often thought, empathy is not a complex achievement of our intellect or something we must labor to acquire. Rather, it develops automatically. It's as natural for us to empathize with others as it is to eat, drink, and breathe.

And just as it takes work to hold your breath, it takes a conscious effort to watch unmoved as someone suffers. Physicians and therapists are well acquainted with just how much effort it takes to protect themselves from an overdose of others' misfortune, confronted as they are every day with their patients' suffering. More than a few of them fail to maintain an adequate border around their own psyches.

Even if we witness a catastrophe only on the nightly news and see the despairing faces of victims in a distant part of the world, we want to help. And who was not moved when Steven Spielberg's poor imaginary creature E.T. felt homesick and wanted to phone home?

Fortunately, we share not only in others' suffering, but in their good feelings as well. The American sociologists Nicholas Christakis and James H. Fowler even claim to have calculated that every happy friend raises our own well-being by an average of nine percent (every unhappy friend, however, lowers it by seven percent).[7]

"Something Pulled My Arm Up"

It is impossible for football fans to stay seated when the hometown boys have scored a touchdown and thousands jump to their feet and cheer. As if a magic power emanated from all the others, your legs seem to have a will of their own. Even before you're conscious of moving, you find yourself standing up and shouting. Sometimes others' excitement even exerts its influence in the privacy of our own living room via television. When the cameras pan to the cheering fans, the home viewer jumps off her couch as if spring-loaded, only to notice with some embarrassment that she's all by herself.

We are constantly receptive to the feelings of others. And since the mirroring mechanism in our head imitates what we observe in them, we often automatically copy the facial expressions, gestures, and tone of voice of our interlocutor. The fact that we so readily adopt the signals of others produces sympathy and the urge to help, but it also can lead to experiences like that of my Austrian grandfather. He told me that during the Anschluss in 1938, when the Nazis marched into Austria and incorporated it into the German Reich, he stood in the midst of the cheering crowd and "something simply pulled my arm up." Even my politically conservative but by no means nationalistic grandfather raised his hand in the Hitler salute before he realized what he was doing. Of course, that does not exonerate all the people who kept their arms up for so long. After all, good sense can pull the emergency brake as soon as we become conscious that a foreign force we find repulsive wants to take possession of our body, our mind, and our feelings.

But the fact that we are so often "infected" by others is the price we pay for our ability to learn. A person who was immune to any outside influence would simply be unable to exist. She would fail at the simplest acts of daily life. The only reason we can wash ourselves, tie our shoes, and talk is that we learned how by imitation. It happens fastest when the brain makes the least possible differentiation between what it sees or hears and what it's supposed to do itself. And that's exactly what the mirror neurons do; they are responsible for both our perception of others and our own actions. Too strong an ego can at least compromise the ability to learn when it intrudes between the model and the learner.

To find out whether learning really functions that way, Gallese's colleagues in Parma set up an experiment with students learning to play the guitar. Beginners were supposed to learn new chords a teacher would play for them, while the researchers recorded the learners' brain activity. And indeed, the same neurons that first reacted to the sight of the teacher's fingerings also guided the learners' hands when they played the same chords themselves.[8]

The fact that humans have incomparably more mirror neurons than do monkeys also suggests that without these brain cells, we would hardly be able to learn anything. A chimpanzee has to watch a long time before it can learn to use stones to crack open a nut, something a three-year-old child can learn in a few minutes. And finally, we also find mirror neurons in the brain's language centers. We obviously owe it to them that we can easily imitate the sounds and words we hear. Songbirds learn their songs in the same way. Young canaries and sparrows have mirror neurons in their tiny heads and listen to the melodies of their elders as if they were producing them themselves—until they can.[9]

Emotional Susceptibility

Like a yawn or a tune you can't get out of your head after someone else has whistled it, smiles are contagious. Since emotions are communicated by the body, there is a direct path from imitation to sympathy, an idea which had already occurred to Leonardo da Vinci. By watching and involuntarily mirroring the gestures of others, we also adopt their feelings. Armed with this insight, Leonardo breathed life into the *Last Supper* and the *Mona Lisa*. "The most important consideration of the painter is that the movement of every figure expresses the condition of their spirit, such as yearning, scorn, annoyance, sympathy, and so on. . . . Otherwise, it is not good art."[10] A work is successful when it causes similar feelings in the viewer. "If the picture portrays terror, fear, flight, mourning, weeping, and wailing or pleasure, joy, laughter, and similar conditions, the mind of the viewers should induce their limbs to move so that they think they are in the same situation as the figures in the picture."[11]

Five hundred years after Leonardo wrote these lines, they were confirmed by the discovery of the mirror neurons in our brains. When we see a smiling mouth, those cells order our own facial muscles to smile. And because our brains generally derive feelings from the condition of our bodies, they interpret this impulse to smile as the expression of our own happiness.[12] We don't even have to really change the position of our mouth, for the actual movement is frequently suppressed at a later stage of processing in the brain. But the signal for the emotion is retained: The mere sight of a happy face lifts our spirits.

At such moments we usually don't even know why we feel good, since "catching" an emotion from another happens without help from our conscious mind.[13] Newborns begin to cry when they hear another baby crying. (Years ago, when all newborns were placed in a common room, this effect could set off a real cacophony.) Monkeys also adopt the feelings of others, as the American primatologist Lisa Parr established by showing videos to her charges. Some videotapes were about chimpanzees happily at play, while others showed scenes of monkeys getting injections or being shot with tranquilizer darts. Animals that have grown up in captivity know and hate the latter situations. The watching chimps reacted as if it were happening to them. Sensors showed that when they watched the threatening scenes, their skin temperature dropped—their equivalent of a cold sweat.

Even the much less intelligent rhesus monkeys are susceptible to the emotions of others. In one experiment, behavioral researchers planned to reward the animals with food whenever the monkeys gave other monkeys harmless but unpleasant electrical shocks. The shockers were able to hear the shrieks of the shockees. The experiment had to be terminated because one of the monkeys preferred to do without the food for twelve days rather than cause another monkey—familiar to him—to suffer the shocks.[14]

But what advantage is there in this impulse to share the pleasant and unpleasant feelings of others? This question probably never consciously arose during the development of primates: The characteristic was simply there at some point. The emotional impulse resulted from the fact that primates in general and humans in particular are extremely good learners. And if we can "infect" each other with gestures and behavior, it was almost inevitable that we could infect each other with feelings as well.[15]

The usefulness of this trait was quickly obvious. If an animal hears the terrified screams of another and itself reacts with fear, then it may escape in time not to end up in the belly of a predator. And if a little girl has unpleasant feelings when she sees a playmate get stung when he touches some nettles, then she probably won't want to try it out for herself. Evolution doesn't ask if a solution is the most efficient; it simply preserves what works.

The prerequisite for emotional susceptibility, however, is that our perception of our own feelings is in good working order. Otherwise, even the smartest of us couldn't share the feelings of others. And that's what happens to people when a head injury has affected the regions of their brain that oversee the body. They cannot feel pain and remain unmoved by a film of someone in extreme agony.[16]

Blind Spots in the Mirror

But as long as our self-awareness is intact, we don't even need to see a face contorted in pain. Our stomach turns over if a friend merely tells us how she banged her thumbnail with a hammer while trying to hang a picture. We're even more affected if she shows us the blackened nail. We are quite capable of imagining what she must have felt and making the experience of the hammer hitting her thumb our own. This feeling is more complicated and far-reaching than emotional susceptibility. It is what we call empathy. In this hypothetical case, we do not experience our friend's pain as if our own thumb had been hit. Our feeling remains abstract in a curious way. We feel uncomfortable, but the trigger—the physical pain itself—is not present. The reaction in our head is also incomplete in a certain sense. A feeling such as pain consists of various components for which different areas of the brain are responsible. Thus one circuit evaluates whether an event is pleasant or unpleasant, another recognizes which parts of the body have been affected, and a third center provides the feeling of pain itself. Finally, the brain assembles these signals and allows us to tell a headache from a leg cramp.

But when we empathize, only the first of those circuits is activated, creating a diffuse unpleasant feeling. The other components are absent, as the Zurich neuropsychologist Tania Singer has been able to show.[17] In the brain of the person who sees or hears about the pain of another, neither the feeling of physical pain nor the assignment of the emotion to a particular part of the body takes place. Thus when we empathize with someone, we're not experiencing an exact copy of what that person feels. The mirror has blind spots.

That distinguishes empathy from emotional susceptibility. If a movie makes us cry when the heroine cries, we are at that moment feeling the same grief she does, as if the same disaster had happened to us. If we have empathy with her, we can understand her feelings but we know her pain is not ours.

The rationale for empathy is obvious: It makes it easier to understand others. For where conscious thought takes detours and often ends up on the wrong track, the gift of empathy grants direct access to another person's interior world. Someone who has never lived with children, for example, may have an abstract idea of what a father must feel when his little daughter learns to ride a bicycle. But only someone who has had the experience himself comprehends the father's feelings concretely. You don't need to explain to him that strange mixture of joy, a little fear, and enormous pride. Few words are needed; you look at each other and you understand what the other is thinking.

Men without Nerves

But whether they come from empathy or emotional susceptibility, feelings are communicated automatically. We are incapable of deciding not to cry in *Gone with the Wind*. And we are involuntarily shocked when we see the photograph of a badly injured person. But just as my grandfather was capable of suppressing (and hopefully did suppress) the spontaneously triggered Hitler salute, we often must stifle our feelings in the face of others' misery, or we would not be able to bear the suffering of the world.

We are capable of regulating emotions that occur automatically, to admit them or close ourselves off to them. The closer we are to others and the more sympathetic we find them, the more likely we are to open our hearts. Experienced hitchhikers look directly into the eyes of approaching motorists in order to soften their hearts. The American social psychologist Mark Snyder found that by doing so, they double their chances of being picked up.[18] And anyone who has found himself in the same situation as a person who is seeking help is more likely to sympathize with him. This is one of the reasons for Vittorio Gallese's feelings as a prison doctor: Being together under lock and key created a bond.

A frightening way to experience the various stages of empathy is to watch the videos of the execution of Nicolae Ceaușescu or Saddam Hussein on the Internet. The scenes are so oppressive it feels like we are being executed ourselves. But as gruesome as the tyrants' end may be, not many viewers will feel sorry for them. The knowledge of all the people they had murdered keeps us from being touched by the images of their own violent deaths.

Who's going to sympathize with a monster when something bad happens to him? Men don't, at any rate, as we learn from another study by Singer.[19] When men learned that someone who has treated them unfairly is getting a very unpleasant electric shock, their brains show almost no pain reaction. On the contrary, synapses concerned with feelings of pleasure are activated, as though the men were feeling schadenfreude. Apparently it is different with women. Their brains

SURVIVAL OF THE NICEST

signal pity even for individuals they have had bad experiences with in the past. No one has yet adequately explained this difference between the sexes. Nor is there any evidence that women *act* any more help-fully or selflessly than men.

Samaritans in a Hurry

Even people who deal professionally with morality are capable of turn-ing off—or even not noticing—their empathy at crucial moments. The American social psychologist and theologian David Batson conducted an experiment that leads one to question the moral effects of religion in everyday life. At the same time, it confirms a central parable of the Gospel story.

Batson asked forty students at the Princeton Theological Seminary to analyze the parable of the Good Samaritan. A man on his way from Jerusalem to Jericho is attacked, beaten, and robbed, and lies half dead in the road. A priest comes by, sees him, and continues on his way. Obviously he fears making himself unclean by touching the corpse of a stranger, for in Mosaic law, priests may touch only the corpses of their own relatives. Next comes a Levite, a scholar of Jewish law, who also passes by without helping. Finally, a Samaritan, the member of a de-spised minority, comes upon the man. He feels compassion and acts accordingly: He washes and bandages the man's wounds, takes him to an inn, and pays for his care.

The students were to prepare oral presentations on this parable in which Jesus illustrates the principle of love for one's fellow humans. On their way to class to give the presentations, the students encountered exactly what is supposed to have once happened between Jerusalem and Jericho. A man sat on a stoop, hunched over and racked with pain. (The students did not know he was an actor only feigning distress.) How did they react? Exactly as described in the Gospel of Luke: They passed by without coming to his aid. "Indeed, on several occasions, a seminary student going to give his talk on the parable of the Good Samaritan literally stepped over the victim as he hurried on his way!"

writes Batson.[20] Only sixteen of the forty pastors-to-be offered any help at all. As a control, Batson had some of the subjects prepare a talk about their future professional expectations rather than about the parable of the Good Samaritan and in addition asked about his subjects' religious attitudes. But neither the subject to be discussed nor the strength of their religious beliefs had any effect on whether they helped the victim or not. What made a difference instead was a sobering variable. Students who were given more time by the investigators to get to their presentations were more willing to come to the aid of the victim, but students who were in a hurry almost always ignored him.

The Spirit in the Machine

"The most compassionate person is the best person, the one most likely to display all the social virtues and all manner of unselfish behavior. Thus whoever awakens our sympathy makes us better and more virtuous."[21]

This simple equation was how the Enlightenment dramatist Gotthold Ephraim Lessing (1729–1781) explained why going to the theater to see a tragedy is an ennobling experience. Tragedies "expand our ability to feel sympathy," and thus our moral faculties are also developed.

Was Lessing right? A frequent misunderstanding is the belief that *only* when our inner feelings are affected can we do something for someone else. A poorly paid public defender is undoubtedly acting magnanimously when she gives a passionate closing argument in a murder trial, but the fate of the accused doesn't necessarily have to affect her deeply. She might even be convinced that the prisoner is guilty and find him completely despicable. Nevertheless, she is able to use all her eloquence to argue the man's innocence, even if she does so only because a case that lacks compelling evidence goes against her sense of justice. To be sure, the wise Lessing did not claim that sympathy was the prerequisite for caring for others. Sympathy and willingness to help are two different impulses.

It is important to differentiate sympathy from both emotional susceptibility and empathy. First, empathy can serve not just selfless aims but also terrible ones. Torturers, for example, use exactly their capacity for empathy to torment their victims in especially sadistic ways. Second, we would hardly consider it sympathetic if in an argument we get infected by the anger of our opponent and begin to shout ourselves. Thus, emotional susceptibility and empathy are no guarantee that we will either feel compassion or help others. Both only orient themselves by the apparent feelings of others, and arise only because those emotions are expressed by the body.

But compassion means also taking into account what the other person is *not* showing, for instance—let's say in an argument between spouses—the other's disappointment and wish for a conciliatory gesture. While empathy perceives only the emotive surface of the other, compassion is concerned with understanding another person. In order to look behind the facades of our fellow humans, we try to exchange roles with them in spirit. In an American Indian expression, the person who shows sympathy "walks in another's moccasins." Lessing speaks of the "fear that we ourselves could become the object of sympathy." For in trying to imagine how we would feel in someone else's place, we must not give too much importance to that person's visible feelings, which threaten to "infect" us. Only by controlling our emotional susceptibility are we able to calm others' fury or comfort them in their despair.

Adopting a stranger's point of view in our mind is consequently carried out by different centers in the brain than those controlling empathy. Among others, an area called the posterior superior temporal cortex (pSTC) contributes to our ability to think our way into others' minds. Its original task was much more elementary, namely, to predict complicated movements of objects. A cat uses it to guess in what direction a mouse will run. In interactions with others, it is especially the folds of our brain behind the right temple that carry out a similar but more abstract assignment. They become active whenever other people appear to be acting in a purposeful and goal-oriented way. Thus with the help of the pSTC, we assign meaning to our observations of what other people are doing. Thus we owe our insights into the inner life of others neither to our conscious reason nor to our feelings. Rather, we acquire them intuitively.

The American neuroscientist Scott Huettel has studied the connection between selflessness and this intuitive understanding.[22] He gave his subjects a task that seemed to have nothing at all to do with sympathy for one's fellows. They were supposed to watch a computer play a video game against itself. The knowledge that the winnings would be donated to a good cause was apparently enough for the subjects to attribute goal-directed action to the machine. At any rate, the pSTC was activated. (In earlier rounds, when the subjects tried the game themselves or when only points rather than money were at stake, this area of the brain was not involved.) Some subjects seemed more inclined than others to see intelligence in the machine, and their pSTC correspondingly showed more than average activity. And it was precisely these subjects who showed themselves to have a particularly altruistic attitude in other tests.

How much people devote themselves to others would accordingly have less to do with their emotional empathy than with their intellectual ability to interpret the intentions and motives of others. This suggests a surprising corollary: The person who has difficulty understanding others is condemned to egocentrism. Looking only to your own advantage, then, would be a form of not just emotional but also intellectual narrowness.

Of course, empathy and the intellectual exchange of roles are not mutually exclusive. On the contrary, they complement each other. Empathy provides a faster and more precise image of the inner life of others; change of perspective shows us sides of them to which empathy is blind. And thus the ability to project ourselves onto other people consists of both things together.

Know Thyself

If you want to put yourself in someone else's role, you must first know what your own role is. We need an idea of ourselves in order to understand others. Just as the mirror neurons replay other peoples' gestures and feelings in our own head, other brain systems reflect others' beliefs and wishes. Parts of the so-called medial prefrontal cortex, right behind the forehead, are primarily responsible for this change in perspective.[23]

However, it is anything but assured that everyone knows what his or her own point of view is. Infants and most animals don't even know who they are. Even in young children and the great apes, the frontal lobe is not nearly as developed as in adult *Homo sapiens*. Correspondingly, the former have at best a vague idea of themselves. That's why it's so fascinating to investigate the selfless actions of apes and very young humans. They provide information about how much self-awareness is necessary to be selfless. And if even little children are ready to do something for others spontaneously and without expectation of reward, it can hardly be because they have been taught to do so. It must mean that at least part of our essential being is naturally concerned for others.

Researchers usually use a mirror test to establish whether a human infant or an animal possesses self-awareness. They dab some color onto the subject's forehead without it noticing, then place it in front of a mirror. If the child or the monkey reaches for his forehead in surprise, he knows that he is himself the figure in the mirror. Until they are about eighteen months old, human children behave as if they were seeing some other child.

But after that, they say their own name when they look in the mirror and try to rub the color off their forehead. And only children who react to their reflection that way are concerned about a playmate whose plastic spoon breaks or whose teddy bear is damaged. That was the conclusion reached by the Munich psychologist Doris Bischof-Köhler in a study of 126 girls and boys and their self-awareness and readiness to be helpful.[24] All the children who helped each other had previously recognized themselves in the mirror, and all the children who did not recognize their own reflection were later unmoved or confused when a playmate suffered a small mishap.

The readiness of small children to help was also investigated by Felix Warneken and Michael Tomasello from the Max Planck Institute for

Evolutionary Anthropology in Leipzig.[25] Their little subjects, all just eighteen months old, watched an adult drop a felt-tip marker on the floor and not be able to reach it, or they watched an adult with her hands full trying to open a closet door that was slightly ajar. The adult betrayed no emotion that might have infected the child. She simply stood there helpless in both situations. And almost always, the children left their toys and picked up the marker or opened the door for the adult. However, if they saw that the adult could take care of the situation by herself, they kept playing. They had obviously understood the problem the adult was having. A wish for reward seemed not to be involved. Quite the contrary: When the investigators began to reward the help with an interesting toy, the children soon became *less* helpful than others who had never received a reward. Apparently, children develop on their own an inclination to help others.

Chimpanzees also recognize themselves in the mirror and behave as if they could comprehend certain intentions of their fellow chimps. For example, they know which food caches another animal is aware of and thus can steal from and which are safe from their rival's raids.[26]

Indeed, there are scattered reports of chimpanzees helping other species. One such comes from the Netherlands-born primatologist Frans de Waal.[27] In an English zoo that had pygmy chimpanzees, or bonobos, a starling once flew into the glass side of their enclosure. A female bonobo named Kuni took charge of the fallen bird and set it on its feet again. She obviously understood the distress of this being so foreign to her, for when the bird did not take off, Kuni picked it up, climbed to the top of the highest tree, and unfolded its wings. Then she threw the starling into the air—which of course resulted in a crash landing. But Kuni was again ready to help. She jumped down from her tree, took up a post next to the poor bird, and protected it from her curious fellow chimps until it finally recovered and flew off.

When Warneken and Tomasello administered the same test they had given to young children to chimpanzees from the Leipzig zoo, their results were mixed. When an investigator dropped his marker, the chimps had no problem picking it up for him. But when he stood in front of the closet door with his hands full, they showed no interest.

Did they not want to help, or were they unable to? In follow-up experiments, the chimpanzees were quite willing to hand things to humans, even when they had to exert themselves to do so.[28] Moreover, they gave fellow chimps access to food even when there was nothing in it for them. However, the difficulties the other was having had to be made very clearly discernible. A person with his hands full standing in front of a closet door was clearly too much of a challenge for the chimpanzees' powers of deduction.

Thus a certain inclination to do things for strangers without expectation of reward seems to also be innate to our nearest relatives. And when they refuse to help, it's not necessarily from selfishness but because they cannot comprehend the other's distress. Helpfulness requires a high degree of intelligence.

The Nobel Prize for Empathy

The chemist and poet Roald Hoffmann opened my eyes to how much depends on understanding our fellow humans even in the "hard" sciences. He is among the most successful chemists in the world, a field where advancement is supposedly based only on intelligence and determination—and admittedly a bit of good luck as well. Hoffmann is the author of more than five hundred publications on the theory of chemical reactions. Universities all over the world have conferred upon him no fewer than thirty honorary doctorates, and last but not least, he was awarded the Nobel Prize for Chemistry in 1981, at the age of only forty-four. Moreover, he has written three plays and four critically acclaimed books of poetry.

When I met Hoffmann at his institute at Cornell University, I asked him what he attributed his successful career to. I expected him to say something about creativity and analytical ability. Hoffmann laughed and explained that Nobel laureates aren't any smarter than other people. That was his experience, and he had attended innumerable conferences with prominent researchers. Curiosity was clearly indispensable, but many people were curious. The difference between him and many of his colleagues was only his talent for empathy. Thanks to it, he had accomplished much not only as a writer but also as a scientist. "I've always had a really good sense of what difficulties my colleagues in the lab are facing—even if they haven't verbalized them. And I've then solved those particular problems."[29]

Hoffmann traces his sensitivity to his childhood in German-occupied Poland, where he survived the Holocaust hiding with his family in the attic of a village school. (Sadly, his father was later shot by the Nazis.) It was impossible to thrash out conflicts in their hiding place. He learned not to cry, for the slightest sound could have given them away. The fear they suffered together during their years in hiding did not merely create a bond among them. It must also have taught Hoffmann, who was only seven when they were liberated, how to interpret the slightest signals from the others.

Of course, Hoffmann is being unduly modest about his analytical skills. But he's right: Many people are intelligent. And his own explanation of his success seems quite plausible in light of the newest research into empathy. Whenever players in the trust game opt for cooperation that is beneficial for both players, the medial prefrontal cortex is activated—exactly the region of the frontal lobe that enables us to experience other people's emotions.[30] Empathy is obviously not just a lubricant for our relationships, but also the royal road to success.

The Empathetic Brain

Since we are used to seeing life as a struggle with others, we easily overlook how often there is nothing to be gained by competition. In such situations, unity is what counts. Just agreeing on a common course of action is often more important than which path we decide to take. When two colleagues are working on the same problem, they will find the solution sooner if they work together than if each one works alone.

The strategic task is not to gain the greatest possible advantage for yourself at the cost of your colleague, but to coordinate your efforts with her. When we think in that way, different processes are taking place in our brain, and thus different synapses are activated. The Taiwanese neuroeconomist Chen-Ying Huang investigated these differences systematically by scanning the brains of her subjects while they carried out tasks from the repertoire of game theory.[31] Huang writes of two different systems in the brain that make strategic decisions. One depends on the dry processing of facts, the other on empathy.

If cooperation is at stake, rational consideration of our own benefit does not necessarily lead us to a useful conclusion. The brain provides us with a better alternative: By taking over the emotions of others, feeling our way into them, and intellectually adopting their point of view, we are much more likely to find a constructive solution.

Whenever the empathetic brain takes over, private intentions are replaced by a common goal—a goal that belongs completely neither to one party nor the other, but stands between them and binds them together. The border between "you" and "me" becomes porous. To a degree, our feelings and thoughts meld together. We see ourselves in the other and see the world through his eyes, but still retain our own point of view. The brain seems to be able to deal with this contradiction by allowing some but not all synapses to be "infected" by the other's feelings and thoughts. And our own and the other's perspective can shift back and forth, as in a reversible figure. In one moment our feelings meld with those of another; in the next we are ourselves again.

CHAPTER 5

There Is Only One Love

The heart has its reasons, which reason does not know.
BLAISE PASCAL, *PENSÉES*

A MORNING WALK IN THE RAIN MAKES us optimistic, because it shows what affectionate beings we are. At every other lamppost or hydrant there's a person with a dog. The owner waits patiently until her companion has done his business. Some owners are talking to their pets, too. Mostly one hears loving words, sometimes admonishment, but seldom scolding. Then human and beast set off for home until the next time—that same evening at the latest—when the dogs again cause their owners to leave whatever they're doing and take them out into the wind and weather.

Almost everyone you encounter this way has a deep relationship to their animal. They call the dog a member of the family (48 percent of all U.S. dog owners said so in a survey), carry a photo of the pooch in their wallet (67 percent), let it sleep in their bedroom (73 percent), and celebrate its birthday (40 percent).[1] And they know themselves to be in good company. In Germany alone, ten million people live with a dog. More than twelve million feel attached to a cat, and another ten million—presumably somewhat less intensely—to a guinea pig, canary, or some other small animal.

How strong many people feel about their pets becomes evident when the animals die. The owners often go through a period of grieving no different than if they had lost a beloved spouse or child. Some become so despondent they can no longer cope with their daily lives.[2] Even the normally stoic Prince Charles was deeply distraught when he had to have his Jack Russell terrier Tigga put to sleep after eighteen years together. Is it any wonder, then, that in 2012, Americans spent a staggering $53 billion on their pets? That's the equivalent of the entire economic output of a medium-size African country. Most of the money goes into the pockets of pet food manufacturers, but veterinarians and more exotic consultants like animal psychologists, homoeopaths, and Bach flower therapists also rake in the profits.[3]

Sometimes people will even risk their lives to rescue an animal in danger. One was the journalist Michael Holzach, famous for hiking across Germany in the 1980s without a penny in his pocket. When his loyal companion, the boxer-mix Feldmann, fell into the Emscher River near Dortmund and was unable to swim back to shore, Holzach jumped in after him. The powerful current smashed the journalist's head against a bridge piling and he drowned. Firemen with a ladder rescued his dog.

From the standpoint of conventional evolutionary theory, it's difficult to explain this common, widespread devotion to pets. After all, people assume significant costs for beings who are of no biological use to them. The lapdog, the cat, and the guinea pig have nothing to give their proud owners that would increase their chances of passing on their genes. At best, the animals lessen those chances by siphoning off some of the owners' resources. According to the traditional, narrow interpretation of Darwinian theory, such behavior ought not to occur.

However, if we can succeed in reconciling the owners' sacrifices with a more inclusive understanding of evolutionary theory, we can also better understand altruistic behavior among humans. We therefore need to examine the forces that bind us to other species. They obviously constitute a motive for selflessness that is stronger than the hope of benefit, stronger even than empathy and sympathy. What are these forces, and when do they express themselves?

Perhaps eavesdropping on the morning conversations of the dog walkers can solve the riddle. Those early risers don't talk to their animals the way we talk to another adult. Their usual tone is the singsong we use when talking to young children. And when the dog owners talk to one another, they very frequently refer to their companions as "this little guy," or "my girl," or "my baby," even when they have a Great Dane on the other end of the leash.

Blood is Thicker than Water

From a biological perspective, it's no surprise that our love of animals resembles our love of children. After all, evolution turns on reproduction. When animals care for others, it must first of all benefit their progeny. That's why there's no question from an evolutionary point of view that there is and must be altruism toward one's own children. Whoever makes sacrifices for his sons and daughters is providing his offspring—and thereby his own genes—a better chance of survival. Even in humans, this motive must have even deeper roots than the wish for material gain.

As a rule, the more highly developed an animal is, the longer its dependence on its parents lasts, and the more time it will later invest in its own offspring. A female sea turtle crawls onto a beach, digs a pit, lays her eggs, covers them with sand, and disappears back into the vastness of the sea, never to see her children. Humans, however, usually care for their daughters and sons at least until the children reach puberty. Nowadays, many a mother still washes her son's socks while he works on his PhD.

But we feel a connection not just to our immediate progeny, since our own genes live on in other relatives as well. That was how the geneticist J. B. S. Haldane, whom we encountered in our first chapter, explained his willingness to jump into a freezing river to save two brothers or eight cousins. To recapitulate the calculation behind his witty remark: Since we all inherit half our genetic material from our mother and half from our father, statistically 50 percent of your genes are in the cells of your children and your siblings, 25 percent in your nieces and nephews, and so on. If Haldane saves two brothers or eight cousins but himself drowns in the act, there's still hope for the survival of his genome.[4]

The German psychologists Franz J. Neyer and Frieder R. Lang have demonstrated that there's a good deal of truth to Haldane's ironic remark. They asked more than 1,300 residents of Berlin whom they felt close to, whom they trusted, and who had supported them in their lives. Their answers showed exactly the gradations one would expect from the genetic proximity of relatedness. The interviewees felt closer to their siblings than to their nieces, nephews, and grandchildren, but closer to the latter than to their cousins, and closer to their cousins than to their friends. Their statements about who had given them support showed similar results.[5]

The findings of the South African economist Dorrit Posel were even more impressive. She analyzed the money transfers from migratory workers to their home villages.[6] The closer they were related to someone, the more they sent them—and the better were the reproductive chances of the receiver as a consequence. Thus, even the inhabitants of a modern metropolis act no differently toward their clan than do animals in the wild. Nightingales feed other's chicks more readily the closer they are related to them. Gophers warn their nearest relatives first when a weasel is approaching.[7]

The Altruistic Gene

The willingness of individual ants, bees, and wasps to sacrifice themselves is also explained by their singular kinship ties. Among these insect species, only the females carry genes from both parents. The males must be satisfied with only the half-set of chromosomes they inherit from their mothers. At conception, moreover, the father gives all his genes to his progeny, but the mother only half of hers. Thus sisters are more closely related to each other than to their own children. They have 75 percent of their genes in common, but only pass on 50 percent to their direct descendants. To ensure the best possible survival chances for her genetic makeup, therefore, an ant would do better to support her sisters rather than her children—or to forego the exhausting mating process entirely. And that's just how these insects behave. Among ants and social bees and wasps, only the queen reproduces. All other females subordinate themselves to the good of the community.[8]

The California psychologist Nancy L. Segal has shown how precisely genetic consanguinity also determines the behavior of human beings.[9] She asked fraternal and identical twins to solve the prisoner's dilemma—the game in which the subjects win when they work together but a single player can do even better by betraying his credulous partner. The twins were able to see each other during the game. If the game is played for several rounds, as it was in Segal's experiment, the partners are able to learn that in the long run, both do better if they trust each other. But the identical twins had more success than the fraternal twins. The degree of their relatedness explains why. Identical twins have the identical genome; fraternal twins are no more closely related than ordinary siblings.

Let us now suppose that there is a gene for altruism, and that in the same way that one person has blue eyes and another brown, some individuals carry the altruism gene and some do not. In reality, of course, things are more complicated. Probably there is not just one gene, but many that are responsible for altruistic behavior, and doubtless the environment plays a role as well. Nevertheless, this simplification allows us to perceive the connection between kinship and altruism. For if a person carries the gene for altruism, the closer someone is related to him, the more likely that person is to have it as well. Whoever has the gene behaves magnanimously. If he meets with a kinsman who has the same predisposition, his magnanimity will be reciprocated. Both will profit, and since they will both then improve their chances of reproduction, the gene for altruism has a good chance of spreading.

That is exactly what a natural law named after the English geneticist William D. Hamilton says: Evolution favors altruists who extend their generosity to creatures with genomes as closely related to their own as possible.

The Brotherhood of Humanity

It's not always that easy to do the right person a favor, however. Anyone who has a large extended family knows how hard it can be to keep all the distant cousins, grand nephews, and great-great-aunts straight. It's often necessary to consult the family tree or ask your omniscient grandmother. Animals, however, have no access to such resources, and even for our prehistoric ancestors, precise family relationships must not have been clear to most people. Under these circumstances, another way must be found to do good things for individuals with genes related to your own.

For instance, how can ants, bees, and wasps—the extreme altruists among the animals—be sure that the queen for whom they sacrifice themselves is really related to them? Obviously, the question doesn't even occur to them. They are simply programmed to blindly nourish the progeny of the queen. Under normal circumstances, they are benefitting their own genes by doing so, since as a rule the queen is their mother or their sister. If one joins a foreign colony of paper wasps to a local one, however, the female workers of the established group will tear the foreign queen to pieces, but her young female workers will be integrated into the established colony and then industriously help raise young to whom they are not in any way related. And since among paper wasps—in contrast to honeybees, for example—basically any adult female can become the queen, that status is sometimes assumed by one of the new female workers. Then the entire established population is working for a foreign ruler.[10]

The wasps are acting logically, for altruism is a question of costs and benefits. And should a selfless act benefit its recipient much more than it costs the giver, it is worthwhile not to draw the circle of receivers too narrowly. Exaggerated stinginess is risky. After all, you never knows exactly onto what soil a good deed will fall. It would be irrational to withhold reproductive chances from a potential carrier of your own genes if being generous doesn't cost very much. It's preferable to accept the possibility of sometimes benefitting someone not related to you.

Birds, for example, behave precisely according to this calculation when they raise a cuckoo's child. The parasitic bird profits from the absolute dependence of the legitimate chicks. Since their own chicks must be fed under all circumstances, the parent birds instinctively feed every open mouth that makes the right sounds. In fact, the longer a particular species' chicks are dependent on their parents, the more willing the latter are to feed foreign chicks as well. Some species, such as the American cardinal, have even been photographed putting worms into the open mouths of fishes.[11]

Homo sapiens isn't too picky about ensuring a genetic relationship, either; otherwise how could you explain the fact that we are more ready to trust people the more they look like us? The British psychologist Lisa DeBruine determined this when she showed her subjects who were playing the trust game manipulated portrait photos of their supposed partners.[12] Apparently, the more someone's features resemble ours, the more we see that person as a near relative.

Whenever people behave selflessly toward strangers, they are transferring a behavior that is of biologic benefit within their family to people outside the family. That is why all cultures describe love of one's fellow humans with kinship concepts: Men become blood brothers, African-Americans call each other "Brother" and "Sister," and children are taught to call friends of their parents "Aunt" or "Uncle." It isn't just the chorus in Beethoven's Ninth who rejoice that *"alle Menschen werden Brüder"* (all people become brothers)—all world religions describe the Kingdom of Heaven as a state in which all people treat one another as siblings.

The Chemistry of Love

Thus in the end, our affection for strangers is also based on the mechanisms of reproduction, but it would be a basic misunderstanding to consider this finding from brain research disillusioning. In reality, neuroscience has seldom achieved a more poetic insight: There is only one kind of love. And this power that binds humans together displays thousands of variations: in sexual desire and in parents' love for their children, in an elderly couple's long-standing intimacy, in friendship, and in our generosity to people we have never seen before. It includes as well our devotion to creatures of other species; it's there when you scratch your dog behind his ear.

Although love may cause endless complications in our psyches, it's fairly easy to explain how it works in our bodies. The mechanisms have been in place for hundreds of millions of years. If you give a male reptile a neurohormone called vasotocin, he will immediately try to mount every female in sight. Thus a single natural neurotransmitter is enough to initiate a fairly complicated behavior. And since evolution has always been efficient in its processes, vasotocin can also do other things. When a female sea turtle crawls onto a beach, scrapes a depression in the sand, and lays her eggs, her blood has elevated levels of the hormone. After she has done her duty as a mother, pushed sand over the eggs, and returned to the deep, the vasotocin concentration declines again. Sex and whatever scant attention to offspring the turtle is capable of are inextricably linked.[13]

It is no different with mammals, including humans. Perhaps because the sexual life of female mammals is more demanding (a female reptile reflexively remains still when the male rubs against her), in the higher animals vasotocin occurs in two variants with minimal chemical differences. The predominantly male variant is called vasopressin and the predominantly female variant is oxytocin, with which we are already familiar as the neurotransmitter for trust. Both are produced in a region of the diencephalon, or interbrain, called the hypothalamus, and each plays an important role in sexual intercourse. During foreplay, vasopressin is secreted primarily in the male brain and oxytocin in the female brain. But during orgasm, the latter seems to contribute to the feeling of elation in both sexes.[14]

Another function of this hormone is to promote a mother's care for her progeny. Oxytocin begins to circulate in her brain during labor and in even higher doses when she gives the newborn her breast. Since among other things, oxytocin suppresses the stress reaction, it triggers the warm, relaxed feeling of nursing.[15] Most important, however, it allows the bond between mother and child to develop. If its effect is blocked in a female rat, she treats her newborns just as she would a stranger's—she eats them. Obviously, oxytocin suppresses the aggressive impulse and turns the cannibal into a devoted mother. Under its influence, a mother rat can identify her own children by their scent. And oxytocin fosters a plethora of other caring behaviors. If it is injected into a mammal's central nervous system, the mammal begins to groom the fur of its companions.[16]

The Casanova Molecule

Oxytocin and its male counterpart vasopressin do much more than just keep mother and child together. Mice provided the first evidence that these hormones are in play whenever an affinity is felt between mammals. Almost all mice are polygamous, that is, they care nothing for their partners once copulation is complete. There is, however, a species in North America—the prairie vole (*Microtus ochrogaster*)—that rivals the most monogamous of humans. As soon as they are sexually mature, these touching little creatures find a partner for life. The male has intercourse with the first female he encounters, and the mating can last for many hours. From that day on, nothing can separate the two. They build a nest, the father cares for their children, and the partners even defend each other against intruders. Their faithfulness outlasts even death: If one partner dies, the other remains alone for the rest of its life.

Their bond is forged during the first intercourse and reinforced during every subsequent mating: Vasopressin is released in the brain of the male and oxytocin in that of the female. This was proved by the American neuroscientist Thomas Insel when he switched off the two love hormones in lab animals. After giving the voles a blocking agent, even the most active intercourse had no consequences; the partners seemed to simply forget each other. When Insel brought them into contact again, they sniffed each other as if they had never met. Their social memory was clearly unable to function without oxytocin and vasopressin.

Even dyed-in-the-wool Casanovas can be transformed into faithful spouses with vasopressin. Insel implanted ordinary, polygamous male mice with a gene from the prairie vole that modified their vasopressin receptors. A receptor is a molecule that receives the hormone on the surface of the gray matter. After the implantation, the mice never left the side of their first love.[17] Insel had similar success with equally promiscuous female house mice by implanting them with a gene for the oxytocin receptor of the prairie vole. A single gene can suffice to forge a bond for life.

Whoever thinks that *Homo sapiens* is above such simplistic mechanisms is mistaken. Like various species of mice, male humans behave differently depending on which vasopressin receptor they have. Some variations in the receptor are more conducive to faithfulness, others to sexual adventurism. In a large-scale study of 522 couples, Swedish investigators concluded that men whose vasopressin receptor was similar to that of common mice complained twice as often of problems in their relationships and could bring themselves to get married only half as often as those with a receptor like that of the prairie vole. The female partners of the latter group were also more satisfied with their spouses.[18] And although no one has yet studied the extent to which a happy marriage depends on variations of the female oxytocin receptor, it has at least been shown that women with a specific variant of the molecule have children earlier than others.[19]

The most important thing vis-à-vis our life together is the fundamental significance of these findings about our neurochemistry: Until a lasting bond between mother and father arose in the course of evolution, only blood relatives were able to benefit one another with selfless actions. But once such a bond appeared, men and women unrelated to each other began to enter into long-term relationships and thus to make their fate and their reproductive success dependent on each other. And that meant that for the first time in natural history, love triumphed over the restrictions of blood relationship.

The Harmonious Hormone

Poets and philosophers often accuse *Homo sapiens* of being the cruelest creature in nature, but in reality, we are the most loving of all. For there are no other animals who care for anywhere near as many of their own species as we do. We bond not just with the partner with whom we intend to reproduce ourselves, but also with distant relatives, friends, and colleagues. When we hear of the distress of distant people we have never met, we even donate money to help them.

Not a single act of affection or care, however, would be possible without those mechanisms that also, for example, drive reptiles to reproduce. Oxytocin and vasopressin, the hormones of bonding and sexuality, are always at play when humans behave cooperatively and altruistically. They promote social relations in general.

In animals, oxytocin can suppress reflexes of aggression or flight. In humans, it promotes trust. That is what happens, for example, when a child approaches a stranger in a trusting way. The experimental subjects in Chapter 3 who sniffed oxytocin prove, however, that this hormone also gives us an abstract expectation of goodwill from others. In the trust game, they were prepared to grant their partner significantly greater sums than other players were.

Contrary to what one might expect, humans high on the love hormone do not see the world through rose-colored glasses. When researches asked the subjects who had been given oxytocin what they thought of their partners, they were as critical as those who had not sniffed the hormone. Nor does oxytocin increase our willingness to take risks in general; when the subjects given oxytocin played the trust game against a computer, they did not increase their payments. Apparently, the hormone only allows us to accept higher risks with other humans. Under its influence, we simply don't feel it to be so bad when a partner disappoints our hopes.[20]

The reason is that oxytocin dissolves our fear and lessens the stress reaction. Under the influence of this hormone, we perceive the faces of others as more friendly and sympathetic, and we are more willing to forgive them.[21] Subjects under oxytocin were still willing to trust their partner with money even after the latter had repeatedly cheated them.[22] They were as indulgent with a stranger as they would have been with a good friend. Thus oxytocin has effects far beyond its original purpose of establishing a close bond between man and wife, mother and child.

Even between strangers, it creates the preconditions for cooperation, trust, and forgiveness.

Cool Altruists

We also owe some surprising insights into the psychic life of egocentrics and altruists to the oxytocin nasal spray.[23] To differentiate between them, the neuropsychologist Tania Singer first observed how her male subjects divided a sum of money with a stranger. Then she subjected the men to unpleasant but harmless electroshocks on their own hands, and finally, they had to watch their (female) partners being shocked. Meanwhile, Singer was using a tomograph to monitor her subjects' brain activity. She observed in the generous men no more signals of sympathy than in the egocentrics. In both groups, there were men whose brain areas for empathy showed strong activity and others who remained relatively apathetic—another indication that sympathy and selflessness have less to do with each other than previously thought. However, there were definite differences between altruists and egocentrics in how they reacted to electroshocks to their *own* hands. While there was strong brain activity in the more stingy men, the more generous men had only a weak reaction.

In a second step, all the subjects were given oxytocin nasal spray and the same procedure was followed. Again, the miracle hormone had an effect—but not on the altruists or the apathetic egocentrics, who continued to be unmoved when their female partners experienced unpleasant sensations. The hormone thus seems not to affect our ability to empathize. The egocentrics, however, now accepted having their own hands shocked with less fear and negative feelings.

For centuries, philosophers and economists (and later, evolutionary biologists) have argued that egocentrics' feelings are rational while altruists' are irrational. Could it be that the opposite is really the case? Singer's experiments suggest that humans do not behave greedily or in a miserly way out of a lack of sympathy but out of fear. Everyone knows people who grab as much as they can because they fear the future or think they'll be cheated out of what's coming to them. Conversely, generous people are often distinguished by their insouciance.

The fear-dampening effect of oxytocin makes this understandable. And in addition, that effect is reinforced by the fact that the hormone makes it easier to adopt the perspective of another person. For example, oxytocin sniffers look their partner in the eye more often and, if only for that reason, are better at guessing what the other person is thinking.[24] The person who can better interpret glances and subtle changes in tone of voice or gesture, as described in Chapter 3, has less fear of being cheated and can afford to be generous.

Thus it is no surprise that our inherited complement of oxytocin receptors is connected to our willingness to do things for other people. People who carry certain models of molecules on their gray cells not only share more generously than others in laboratory experiments, but also assert in surveys that other people are especially important to them.[25]

But as always, genes are only one of several factors influencing our behavior. Our life experiences also affect how much we trust others and how readily we give. But the specifics of each situation also play a surprisingly large role in behavior. If we feel uncertain, we mistrust others, even if the feeling of threat has nothing to do with the other person. On the other hand, after a gentle touch—to say nothing of a massage—our oxytocin level rises and with it, our wish to be generous. Experienced waitpersons know that they can increase their tips by seemingly unintentional brushes against their customers while serving their food.

Giving Makes Us Happy

People who give of their own free will exchange possessions for trust—trust and happiness. And the converse is true as well: The more people experience joy in giving, the more they will give. As obvious as it sounds, this contradicts the rumor that unselfishness means self-denial.

That widespread misconception is highly implausible in any case. It assumes that people would act against their own feelings. But brain research of the last two decades has been able to demonstrate how wrong such an idea is. Our emotions are the motive force behind almost all our decisions.[26] Even when we make a dentist appointment to get a filling, it's not really a rational resolution that drives us but the fear of even more pain if the tooth decay continues.

Why would we neglect our emotional interests only in our relations with other people? Neuropsychological experiments, at any rate, came to the opposite conclusion. While subjects were deciding whether and how much they wanted to donate to good causes, investigators measured their brain activity and discovered that at the moment of giving, the same regions of the brain are active that also trigger joy at receiving a gift. The reward system is responsible—that web of gray cells we have already described whose center is the midbrain. It triggers feelings of pleasure whenever a situation seems advantageous. The act of giving thus makes us happy in the same way as good food, an unexpected gift of money, or sex.[27]

However, the joyful excitement is differently orchestrated by the brain. In a decision to do a good deed, signals also reach regions where the binding hormones oxytocin and vasopressin circulate. When we do something for others, the brain prepares to strengthen our relationship to those people.[28] In this way, we form a friendship that may benefit us later. Nevertheless, such intentions are by no means behind every act of generosity, for the responsible circuits on the underside of the cerebrum have little to do with planning for the future. Rather, they reflexively turn on even when we know that the receiver of our generosity will never reciprocate.

We even feel joy if we only see someone getting something from a third person. We certainly can't expect to profit from that, either. We owe our pure happiness at the welfare of others to the reward system. As brain scans show, people have different levels of happiness at others' good fortune. The stronger their brain activity was, the more they were prepared to give something themselves. Thus these people didn't do good because they wanted to play the role of a benefactor. Rather, it was the result of their action that seemed to count: It was a pleasure to them that a fellow human being was happier after the good deed.[29]

Noble Savages?

Such insights touch upon a favorite topic of debate everywhere from cocktail parties to philosophical symposia: Are humans by nature good or evil? Since there's no denying either how vicious or how loveable our species can be, such debates almost inevitably end up in mutually contradictory positions. The pessimistic camp maintains that humans are born egocentrics and come to be socialized only by upbringing and threats of punishment. The optimistic camp, on the other hand, claims that in the state of nature, humans are good and are made bad only by bad company.

Both positions have a long history, but no one has articulated them better than the Enlightenment philosopher Thomas Hobbes and his Geneva colleague Jean-Jacques Rousseau. Hobbes, who lived through the devastating English civil wars of the seventeenth century, saw humankind in a state of nature above all concerned with survival. And since to survive in a crisis, one must sometimes do so at the expense of fellow humans, Hobbes said that one ends up with the "warre of every man against every man." For that reason, life outside the bounds of justice and law would be "solitary, poore, nasty, brutish, and short."[30] The only solution is for humans to submit to a state that will teach them what they do not innately possess: sympathy and the willingness to share.

The optimistic Rousseau, on the other hand, popularized the concept of the noble savage. Humans are basically good, he thought, and sympathy a natural instinct. If it was nevertheless true that conflict and corruption ruled the day, that was because society corrupts us. By placing ourselves in hierarchies and making ourselves dependent on others, we were trying to adjust ourselves to a condition that was counter to our nature. Unfortunately, we are not particularly social beings, thought Rousseau. To escape this dilemma, we would need to develop a form of living together that gave each individual the greatest possible measure of freedom. From then on, idealists who intended to reform humankind by reforming society generally invoked Rousseau.[31]

Parents, however, will probably not find themselves agreeing completely with either Hobbes or Rousseau. They often experience their children as beings with two faces. On the one hand, small children are little savages who bite, kick, and snatch toys away from one another. All the more astonishing, then, how sympathetic and helpful these little beasts can be the next moment. When my wife was sick in bed, my twenty-month-old and rather rowdy daughter suddenly brought her favorite stuffed animals into our bedroom and laid them on her mother's stomach. On the playground, that same physically rough child would sometimes worry about a crying baby. And when she saw a young monkey fall from its perch in the zoo, she cried "Ouch!" and was upset.

As we saw in the previous chapter, small children are quick to be helpful as soon as they understand that someone needs help. In all cultures, children automatically begin to care for others just as they begin to walk and talk. This observation suggests that our social inclinations are not only the result of our training. In this point, Rousseau had the better nose than Hobbes. We are obviously born with a predisposition to care for others.

The Pain of Rejection

Jean-Jacques Rousseau, however, was mistaken in his assumption that it was society that corrupts. The opposite is true. Only because we live together with other people are we even capable of sympathy and sharing. Without the brain systems that are responsible for our attachments to other humans, altruism would be unthinkable. The savage as Rousseau imagined him would not be noble, but selfish.

In the millions of years it took humans to develop from the earliest primates, an attachment to others became a necessity. On his own, *Homo sapiens* cannot survive long in nature. And that's why evolution implanted in his brain the need to be in a community with others of his species. How deeply social feelings are anchored in us is revealed by the common idiom that someone has "hurt us" through disregard. For that's exactly what the person has done. When we feel rejected, the same brain centers in the so-called pain matrix are activated as when our body receives a physical injury. A mechanism with the vital task of protecting our physical well-being thus also reports disturbances in our relation to others. In laboratory experiments, all it took was to shut subjects out of participation in a silly video game to trigger a significant pain reaction.[32]

Luckily, nature conversely rewards the prospect of a successful relationship with good feelings. After all, the reward system not only is activated as soon as we have a friendly encounter with another person; it also entices us to be generous ourselves. In addition, the brain secretes opioids whenever we are in the company of people we feel close to. These chemical twins of opium are manufactured in the pituitary gland and the hypothalamus and allow us to feel the warm happiness of emotional security.[33] What's more, since oxytocin reduces fear and stress, evolution has produced plenty of incentives for us to forego a short-term advantage in favor of attachment to others.

Altruists Live Longer

Living together in harmony makes people happy—and healthy, too. The new discipline of psychoneuroendocrinology is exploring how the hormones that modulate interpersonal communication also affect one's physical well-being. The stress-reducing effects of oxytocin and the opioids have especially positive effects on health. Both diminish the secretion of the stress hormone cortisol and thus protect against not only cardiovascular diseases but also infections, since chronic stress damages blood vessels and also interferes with the immune system.[34]

Such effects possibly solve the riddle of why women in all cultures live longer than men, although pregnancy and childbearing take such an enormous toll on their bodies. As the California neuroscientist John Allman discovered, male apes also die earlier than females, but only among species where the mother raises the young alone. The New World monkeys called titis (the genus *Callicebus*) are the exceptions among primates, for after the young are born, the father takes over their care. The mother only puts in an appearance to nurse them. And these male titis outlive their mates by more than 20 percent. Among the siamang gibbons (*Symphalangus syndactylus*) of the Malaysian rain forests, both parents raise the young together, but their offsprings' attachment to the father is stronger. Here, the lifespan of the males is still 9 percent longer than that of the females.[35]

In the same way, a strong social network increases humans' life expectancy. This is not just a result of the binding hormones, but also because people with stable contacts with others also take better care of themselves. A classic long-range study of more than 6,500 subjects in the San Francisco area showed that intensive social relationships halve the risk of death at any given age. This number seemed so incredible that many researchers set out to reproduce them, and they all achieved similar results.[36]

The decisive factor for longer life expectancy is not that people are supported by good friends—not what we get from others—but what we give to others. This remarkable conclusion was reached by three large long-term studies, one of 1,200 Spanish retirees, and the others of 400 and 1,500 American retirees respectively.[37] The investigators assessed the health of the retirees of both sexes and questioned them in detail about how much help and encouragement they called upon from relatives, friends, or neighbors, as well as how much they themselves helped others. Half a decade later, the investigators compared the interview results with mortality data from the group. The more the elderly subjects took care of others, the more likely it was that they were still alive. It was irrelevant whom they had helped, nor did their own state of health at the beginning of the study play any role. Although it may be difficult for seniors in poor health to go shopping for a neighbor or to look after their grandchildren, among similarly disabled retirees, the ones who helped others despite their infirmities always survived the longest.

To Be Calculating Is to Be a Loser

It's not easy to assess the costs and benefits of a selfless act. Superficially, people who give more than they receive put themselves at a disadvantage. In the long run, however, the willingness to give can pay off. For if people improve their health and life expectancy by helping others, they have thereby also improved their reproductive fitness. Even if they are too old to have children anymore, they still can make it easier for their genes to be perpetuated in their relatives.

Even a dog—a parasite, from a strictly evolutionary biological point of view—can be beneficial to his master just by needing her care. There's a good reason why breeders of many kinds of pets have bred them to have big eyes and turned-up noses like small children. Shouldn't evolution long since have eliminated the delight we take in caring for helpless creatures? At the very least, it would not lessen the reproductive chances of a person if he could secrete beneficial hormones without having to take his dog for a morning walk in the rain to do so. But that is a fairly theoretical idea. The cuckoo, too, has been in existence for millions of years, although the birds into whose nests it smuggles its eggs would clearly have more of their own young survive if they could identify the intruder chick. But to effectively protect themselves from the swindle, they would have to react to open beaks more selectively in general, and that would be too high a price to pay. The bottom line is that tolerating the parasite and accepting the occasional loss of a clutch of their own eggs is the more successful strategy.

It is exactly the same in our ties to others. People who constantly calculate costs and benefits will certainly be fleeced less often. Nevertheless, such individuals will hardly be able to come out on top in the long run. For human existence is much too complex to allow the cost-benefit balance of a relationship to be predicted or even approximated. Compared to almost all other animals, we live longer and are extremely dependent on one another. Moreover, our life together is so flexible that we cannot foresee if a favor will ever be repaid or repaid in excess. And the more complex the social fabric is, the more important it is to join together successfully. Even among the great apes, where conflicts are worked out only in physical battles, the status of a male depends less on his physical strength than on his ability to forge alliances. Among humans, it is even more important that we not be too miserly. Thus *Homo sapiens* achieved evolutionary success as a creature who can quickly and easily form bonds with others—because good feelings seduce us to do so. For emotions are nothing more than aids in making decisions; they allow us to quickly evaluate complex situations.[38]

Characteristics such as amiability, gentleness, and helpfulness developed because they gave their possessors an advantage in evolutionary competition. But that doesn't mean that every single selfless act will bring benefits.[39] No one would claim that sex was an ineffective way to reproduce ourselves just because not every copulation results in a pregnancy. Nor do people jump into the water to rescue a drowning person because they expect eventually to get something in return. The evolutionary advantage of altruism is rather that a creature endowed with selfless impulses tends to reproduce with more success than a calculating rascal. The costs of being altruistic—like the costs of feeding a cuckoo's chick—are bearable as long as they are outweighed by the benefits.

But the greatest advantage of selflessness was that it helped our ancestors to develop larger brains. This step was possible only because our distant progenitors began to share and cooperate like no other creature. To do that, they had to overcome the boundaries of self and learn to see the world with the eyes of another and feel like another. Altruism is what made us human.

PART II

ALL OF US

CHAPTER 6

Humans Share, Animals Don't

TSITSIKAMMA IS WHAT THE LOCALS CALL THE desolate coast where the African landmass plunges into the Southern Ocean. The name means "where the water begins." Standing on these cliffs and looking south, you see nothing but waves that go on for 2,500 miles before breaking on the ice of Antarctica. How many hikers on these rocks guess that beneath their feet is one of the oldest scenes of human history?

To find it, one follows the little Klasies River, which has worn a deep cleft from the highlands down to where it empties into the sea. Descending past several waterfalls, you come to a bay. Unexpectedly, only a few yards above the shoreline, there is the entrance to a system of caves. These caves must have been often submerged, for their floor consists of soft sediment. One can see shell fragments in their walls in addition to curiously chipped stones and bones, as if someone had cemented them in.[1]

By the 1960s, these deposits had begun to attract the first archeologists. It quickly became clear that the bones were from animals killed in the hunt and the stones were blades from the Mesolithic period. Further excavations revealed richer traces of human prehistory than almost anywhere else on earth. The sediments are almost seventy feet deep and contain layer upon layer of artifacts from more than 110,000 years: the ashes of ancient fire pits that tell of the cave-dwellers' diet, traces of ocher that suggest they decorated their bodies. One of most exciting finds was the bones. For the human skeletons that lay even in the deepest layers of sediment were practically identical to the skeletons of humans of the present day. The men and women who lived here more than 100,000 years ago must have looked just like us. The archeologists had found nothing less than the oldest known remains of modern human beings.

Like the pages of a history book, the layers of sediment tell of the life of our early ancestors. The remains of the animals they hunted are especially informative. For example, there were no remains of birds or fish. Apparently, these humans had not yet developed bows and arrows or boats. On the other hand, they had learned how to bring down powerful mammals. Among the bones of antelopes and other small ungulates that were apparently the most important source of meat, the American archeologist Richard Milo discovered the bones of the now-extinct giant buffalo *Pelorovis antiquus*. This animal, which grew to almost ten feet long and weighed more than three tons—as much as a small elephant—must have been pursued by the Mesolithic hunters as well. For under the microscope, Milo found indentations on the buffalo skulls that could only have been made by sharp stone wedges with which the cave-dwellers had belabored their prey, either to kill it or to cut up its body.[2]

There is no doubt that they hunted the enormous animal together. A single hunter could never have bested the buffalo; *Pelorovis antiquus* would have simply trampled any hunter who attacked by himself. Even if the hunters brought down the giant buffalo by luring him into a pit, as Milo conjectures, it is hard to imagine that a single hunter could have dug a big enough hole working alone. We must conclude that even the earliest representatives of modern humankind were capable of cooperating—and thus also of dividing their prey into shares after the hunt.

Besides the development of language, this was our species' all-time greatest achievement. The future history of humankind would be unthinkable without it. If our ancestors had not learned to follow common goals, they would never have become sedentary, never have crossed the oceans and colonized the entire earth, and of course, never have invented music, art, and all the comforts of modern life. To be sure, neither would they have waged wars or brought their home planet to the brink of environmental collapse. Today, we can imagine no other life for ourselves—the circles of friends, the clubs, businesses, and states in which we circulate and live have all become second nature. It never occurs to us what an unlikely thing it is that humans work together in groups.

How did it happen? Part I of this book has detailed how cooperation and even altruism can survive and prosper because they so often pay off in the long run. It also described what goes on in our heads when we trust, feel empathy for, or establish ties with others. But our arguments up to now assume that we are the way we are; that is, that we have the fundamental ability to cooperate. Now we must explore the path our species has followed to achieve this ability.

It Hurts to Share

To live together, we have to share. But we are much more resistant to giving something to someone else than to helping her. One can observe this difference very clearly in children. Even though one-and-a-half-year-olds will support each other in difficult situations, they are by no means willing to surrender their own toys. On the contrary, the little tykes defend their meager possessions with screams and, if necessary, blows. (There was no word I heard more frequently from my daughters when they were still in diapers than "Mine!") It takes at least another year before children are ready to give something away spontaneously. This is not just the anecdotal experience of parents plagued by constant bickering between toddlers; it has been shown by controlled experiments as well.[3]

Even for adults, it's much easier to do a good deed for someone whose need has awakened our sympathy than to concede him a right to our possessions. Quite a few of our well-to-do contemporaries who are touchingly concerned about wildlife protection or disadvantaged children pay their cleaning lady poverty wages and howl about every dollar they have to pay in taxes.

Why is sharing so difficult? Because there are always a few people who want the most for themselves.

When we have to deal with such people as individuals, we can always protect ourselves against being exploited by avoiding them. After all, the result of tit-for-tat is that we are very careful in choosing those with whom to cooperate.

For a whole group, on the other hand, it's much harder to deal with freeloaders. The hunters who left buffalo bones behind in the caves on the Klasies River took the trouble and ran the risks of the hunt. There must have been a big temptation to stay at home and then profit from the others' kill. It's true that in the best case, the efforts of the hunt are still worth it for the hunters, even if some others are shirkers. They will still go out and hunt for their own advantage, and it won't matter to them if the freeloaders get some of the left-over meat.

But more often it's the case that doing battle with a giant buffalo is only worth it when at least the majority take part. If too few participate voluntarily, it's better to go after smaller, less dangerous game. Why should you risk your life so a few good-for-nothings can fill up on buffalo meat? Killing a gazelle will suffice.

And so the community is in the same trap Rousseau analyzed in his story of the stag hunt: For the group, it would be best if as many as possible cooperate, but for the individual, the most rational course is to leave the risk to others and only show up when your fellow clansmen are dividing up the meat. For if the successful hunters also behave rationally, no one will go away empty-handed. After all, the daring hunters and their families can't possibly eat the whole buffalo by themselves. Thus cooperation is rewarding, but it's even more profitable to refuse to cooperate—the prisoner's dilemma once again, but this time with many participants. And that makes the whole thing much more ticklish than the two-person situations analyzed in the first part of the book.

Of course, the hunters could have gotten even with their lazier or more cowardly friends. They would only have to deny them access to the kill. But punitiveness has its costs. For instance, someone would have to stand watch beside the colossal carcass and possibly fight the slackers off. It's just the prisoner's dilemma raised to another level. And can we blame a weary hunter for not wanting to stand guard all night long to defend his principles?

The Tragedy of the Well-Intentioned

This dilemma is universal right down to the present day, and perhaps even more threatening now because the interconnections between human beings are so much more complex than in the days of the hunter-gatherers. Parents at a school agree that the classrooms are in urgent need of a new coat of paint. A few volunteers could easily take care of the job before lunch some Saturday, but nothing gets done because everyone is waiting for other fathers and mothers to volunteer. After all, your own child has the benefit of a more attractive classroom even if you don't touch a paintbrush. Everyone justly complains about the high cost of health insurance, but many are happy to have their doctors order tests that are expensive and unnecessary. Everyone knows we can't keep pumping carbon dioxide into the atmosphere, but no one wants to forgo the comforts of her SUV, especially if her neighbor drives an even bigger one.

Laboratory experiments show with frightening clarity how quickly cooperation can break down even in a group with only a handful of members. (The following chapter will discuss this in more detail.) And when the group gets larger, trust disappears even faster. For with every additional member, it becomes more likely that one person will shirk responsibility. As soon as someone else notices the swindle, that person is tempted to become a shirker, too. And so a domino effect begins. Even someone who was well-intentioned at the beginning can now hardly do anything but put his own interests first.

The American ecologist Garrett Hardin wrote a famous article about this behavior entitled "The Tragedy of the Commons."[4] The title refers to the common pastures and woodland open to all the inhabitants of a village community in the Middle Ages, land that often became mercilessly overused. One could just as well call the problem "The Tragedy of the Well-Intentioned." Even when people want to contribute to the general welfare, they are often unable to do so. Sometimes we even find ourselves forced to knowingly plunder common resources against our own long-term interests because we see that it is happening anyway and it seems better to get something for ourselves rather than leave it all to others.

Are Vampires Altruists?

All the more valuable, then, are the buffalo bones from the Klasies River, for they prove that even the earliest humans found a way out of the trap—without the need of laws or police. And thus, as we know today, they became a unique phenomenon in the natural world.

For a long time, behavioral scientists believed that cooperation also occurred among other species on the pattern of the exchange of favors between Adam and Eve described in Chapter 2. Until just a few years ago, in fact, so-called "reciprocal altruism" among other animals was considered the evolutionary origin of our own selflessness.[5]

But that theory did not hold up under scrutiny. All studies show that both small children and other animals are much more likely to help than to share. Even if chimpanzees comfort one another and pick up a dropped pencil for their keeper, even if dolphins save swimmers from sharks—as soon as food is at issue, the helpfulness of these creatures is at an end and it's every animal for itself. Outside of close relatives, animals very rarely share with others.[6]

Understanding what keeps them from sharing sheds light on how humans live together. For one thing, the basic problems are the same whether what's to be divided up is a bunch of bananas or the right to release carbon dioxide into the atmosphere. For another, comparing other animals to humans gives us insight into how our ability to cooperate must have originated.

The classic example for supposed altruism in the animal kingdom was none other than the not very attractive vampire bat. This New World mammal's diet consists exclusively of blood from other animals that it parasitizes at night. However, since it doesn't find a host every night, its existence is precarious. If it goes without nourishment for more than sixty hours, it will die. On the other hand, if it succeeds in sinking its teeth into a cow, a horse—or even a human—then the vampire drinks so much that it has some to spare. Back in the cave where it sleeps, it vomits up undigested blood.

Almost always, it feeds the extra blood to its children and sometimes other family members as well, as if it were familiar with William D. Hamilton's theory that evolution favors altruism among relatives. Every now and then, however, the vampire bat also bestows some excess blood on bats it is not related to. Three decades ago, this discovery created a furor. The new discipline of sociobiology thought it was on the track of reciprocal altruism among animals, and many textbooks still state as much as fact. Sociobiologists argued that the donors were simply making down payments. If they shared blood today, they could expect to get a transfusion the next time around. The vampires were in effect trading in blood.

But is that what they're really doing? Now that entire theories—even theories of human social interaction—have been built on the bats' nocturnal behavior, it's worthwhile to take a closer look. All these theories are based on a single study published by the American biologist Gerald S. Wilkinson.[7] In the early 1980s, he spent a total of twenty-six months observing the nocturnal behavior of vampire bats living in fourteen different hollow trees on a cattle ranch in Costa Rica. He was able to document 110 acts of vomiting up blood, with an average length of sixty-three seconds. In seventy-seven cases, a mother fed her child. In twenty-one cases, some other relative drank the excess blood. In only twelve cases was the beneficiary an individual that Wilkinson was not able to clearly identify as a relative of the donor. Was the biologist just fooling himself? Did the bats donate the precious blood to a stranger by accident, or was it really a case of reciprocal altruism?

In order to answer these questions, Wilkinson captured eight females, not related to each other, from two separate colonies and got them used to living in captivity. Then each evening, he removed a different animal from the cage and deprived it of food while the others were able to drink blood. Early in the morning, Wilkinson brought the excluded animal back and observed what happened. There were thirteen occurrences of vomiting with an average length of twenty-five seconds. Four times, a blood donation was reciprocated; that is, one bat fed a second bat that had fed the first one the night before. Now Wilkinson could be sure that the donor and the recipient were not related to each other, but did the bats know that, too? Questions are raised by the fact that in twelve of the thirteen cases of donation, animals from the same colony were feeding one another. This suggests that the donor females believed they were helping their own children, nieces, or aunts.

This fairly obvious explanation seems especially plausible because reciprocal altruism assumes a good deal of reasoning. Whenever an individual does a good deed for another in the hopes of receiving some benefit later on, it must first of all forgo something. Anyone who has ever tried to put off eating a piece chocolate until the following day knows how difficult it is even for *Homo sapiens* to postpone pleasure. Young children are completely incapable of deferring gratification, while grown-ups require interest payments to do so. If we can get $100 right away or next year, we want it on the spot; if we agree to wait until next year, we want $105.

Behavioral researchers obtained similar results when they investigated how long animals were willing to wait for a proffered treat. Rats require much higher interest payments than humans, and pigeons require even higher ones. One experiment allowed pigeons to choose by pushing a button whether they would get a certain amount of grain immediately or twice as much two seconds later. Almost all the birds chose the former.[8]

Repayment of a debt on the principle "Give me something and I'll give you something" would soon become so expensive with these animals that cooperation would necessarily break down. What's more, it's unclear that they would even remember that another individual had done them a favor. If they are inherently incapable of doing so, a donor would have no reason to make a down payment. And even if rats and pigeons were to be able to remember, they would have to keep track of expenditures and income—they would have to be able to count. All of this involves complicated intellectual operations.[9]

Thanks to the operations of their frontal lobe, adult humans are capable of self-control and therefore of reciprocal altruism. This most highly developed part of the brain is more prominent in *Homo sapiens* than in any other animal.[10] It is also the region that develops the slowest during maturation. Only after puberty does it reach full functionality. That also explains why two-year-olds want everything at once and won't part with a single jelly bean on the promise that they'll get more later.

Are vampire bats really smarter than toddlers? Scientific progress depends on the skepticism of colleagues who check questionable results by trying to reproduce the experiment. But no one has ever dared to try to reproduce Wilkinson's time-consuming experiment with vampire bats. To be sure, there have been numerous attempts to find other evidence of reciprocal altruism among animals, but all attempts to clearly demonstrate tit-for-tat behavior among jays, lions, or capuchin monkeys have failed.[11] It would be more than astonishing if vampire bats were the only animal besides us capable of exchanging favors.

Lions and Warthogs

Another example that defenders of cooperation among animals like to cite is the hunting behavior of lionesses. It is true that they hunt together, but only when they are pursuing large quarry such as buffaloes or zebras. Only as a group can they bring down such large prey. Once a kill has been made, all the lionesses in the pride gather, as well as their young and of course the male, who seldom bothers to hunt. Now all the members of the pride may help themselves, whether they participated in the hunt or not, and independent of their degree of kinship.[12] Thus, while the hunting females have expended the energy and run all the risks, the shirkers get to eat for free. Everyone benefits from the effort—for one buffalo can feed the entire pride—but some benefit more than others.

On the other hand, if a warthog comes by and a lioness goes after it, the others don't move a muscle to help her, as if they knew that the prey doesn't stand a chance against her. But whoever can get to the kill in time is allowed to eat, because lionesses don't give a damn about fairness. They are completely incapable of caring about justice because they know nothing about the principle of punishment. But since on balance the huntress whose effort is being exploited also profits, the pride is not going to run out of food.

But how can we be sure that early humans were capable of real cooperation and sharing, instead of just following the path of least resistance, like the lions? In the caves along the Klasies River, the remains of roasted penguins and shellfish have also been found. It is food that individuals acting alone could easily have caught, but it must have been shared and eaten together, otherwise it never would have been brought into the cave in the first place.

Even more impressive are the anthropologist Mary C. Stiner's analyses of animal bones from various caves in the Levant.[13] They, too, show marks made by stone tools, marks that reveal that more than 300,000 years ago, early humans cut up the carcass of a slain deer together. Apparently, everyone was allowed to simply help himself, since the nicks on the bones run every which way. But about 100,000 years ago, during the transition to the Mesolithic, the cutting technique suddenly changes. Now it looks as if the meat had been processed by a butcher who divided it in a systematic way. Apparently, by this time, humans had learned to share their nourishment in a more organized fashion.

Lionesses, by contrast, do not even reliably stick together when attacked by an alien male. Normally, such an intruder intends to take over the pride of females. To make room for his own progeny, he will often kill the young of the pride. Together, the lionesses could chase the aggressor away and save their young, but usually they do not. If one of them goes up against the enemy male, the others stand around passively. Over several decades, the American scientist Craig Packer, a world authority on lion behavior, observed the big cats in every corner of Africa. Astonishingly enough, he writes, it is the same females who time and again will risk themselves for the others. He found no trace among lionesses of the female solidarity that has so often been praised.[14]

Grumpy Chimpanzees

It would be astounding, after all, if cats were capable of cooperative action that is not even possible for our nearest relatives in the animal kingdom. It's true that wild chimpanzees in West Africa do conduct regular drives against the smaller colobus monkeys. One chimp chases a colobus up a tree and others then climb into its branches to cut off the smaller animal's escape routes. But what looks like coordinated action turns out not to be so under closer inspection. For the hunt begins with a single male chimpanzee stalking a colobus; only then do others who happen to be nearby join in. Their goal is not to kill their prey in concerted action, however. Instead, they are all trying to profit from the initiative of the first hunter and grab the colobus for themselves. Once one of them has killed the prey, he does everything he can to escape alone with his prize.

But usually the killer is immediately surrounded by a horde of beggars. He's forced to let them all help themselves. The individuals who get the most, however, are not necessarily the ones who helped to hunt the colobus down, but simply those who are the most aggressive eaters.[15] Thus the killer does not act from any sense of justice or even to gain the goodwill of his helpers for the next hunt. He simply gives in to force, since he must fear having to fight for his prey with an especially aggressive beggar.

Chimpanzees basically never share. Chimp mothers are not particularly generous even with their children. If they ever do give their progeny something, it's only when the young beg for it. But in two out of three cases, the mother refuses to give anything. And in the rare cases where she does give them some food, it's always an inferior piece; the child has to be satisfied with the peelings, while the mother gets to enjoy the fruit itself.[16]

Primatologists report that there is simply no provision for sharing in the behavioral repertoire of the great apes, because the animals can feed themselves adequately from leaves and fruit. Each individual is able to find its own nourishment; there is no advantage to be gained from looking for food together. When behavioral researchers experimentally gave chimpanzees food when they pulled on a rope together, the animals only worked together as long as the food was distributed in two separate portions. As soon as it came all in one portion, they hardly touched the rope again—the prospect of not getting any food was too unbearable, and the problem of agreeing together was insurmountable.[17]

Do the chimps fail because they are unable to see the matter from a partner's perspective? In any event, in experiments where sharing is at issue, they behave as if utterly indifferent to the welfare of their partner. When chimpanzees are given the choice of only getting a treat for themselves or getting one for another individual as well, they choose both variations with equal frequency. Thus they have no preference; it is only chance that decides. They are not interested in pleasing a fellow chimp even when they themselves would lose nothing by doing the favor.[18]

In other respects, too, the behavior of the chimpanzees looks suspiciously like that of *Homo economicus*, that perfect egocentric whom some economists see as the ideal image of humanity. For example, this theoretical construct is supposed to be unacquainted with pride, and so, even if someone makes him an outrageously shabby offer that any real-life human would reject out of hand, he will accept it. And that is exactly what chimpanzees do. After all, it's better to get one measly raisin than none at all, even when the donor himself has a whole bag full of raisins. That seems logical, but chimpanzees demonstrate the limits of such petty calculations: It is we, not the chimps, who have conquered the world and flown to the moon.

Rainforest Adoption

While our species has a gigantic head start toward sharing, this is certainly not because chimpanzees are fundamentally not as smart as humans. If two-and-a-half-year-old humans, adult chimps, and adult orangutans perform the same intelligence tests, there is no marked difference in most categories.[19] Whether it's a question of recognizing number and size, cause and effect, or the order of things in space, the apes master these tasks at least as well as small children. (Language played no part in the tests.) And in certain tasks requiring memory, chimpanzees can even do better than adult humans.

It is only in their social intelligence that children prove superior to chimps and orangutans. If girls and boys working together can solve a problem that defeats adult great apes, it is obviously only because they can learn better by watching others, putting themselves in their places, and predicting their behavior.

What is the circumstance to which we owe our social intelligence? The conventional answer is that we are more intelligent and therefore more skillful in dealing with others thanks to our larger brains. But a voluminous brain all by itself doesn't make us smarter. It has often been pointed out that in the ratio of brain mass to body weight, the tiny shrew is better endowed than a human. And although the brain of a two-year-old human already weighs twice as much as a chimpanzee brain, the child is not generally superior to the ape except in its social skills and ability to learn.

Some special circumstance must have led to the development of these abilities in our ancestors. For if high social intelligence were a fundamental advantage, chimpanzees could have developed it, too. But apparently, they had no occasion for it, while early humans did.

We will never learn exactly what happened in Africa more than 100,000 years ago. But the surprising habits of chimpanzees in the Taï National Park in Côte d'Ivoire hold some clues about what conditions promoted human development.

The apes in the Taï forests are under constant threat from a large population of leopards. Sticking together is their only protection against the big cats. By staying united, the life of each individual becomes valuable. If the band of apes shrinks in number, it makes it easier for the predators to get at them. This constant danger presumably explains why the Taï chimpanzees stand up for one another, which they would otherwise not do.[20]

They not only defend themselves together against the leopards, but they also care for members of the troop who have been injured in the fight. The Leipzig primatologist Christophe Boesch observed what amounted to a medical service among the animals. For hours, the chimps licked clean the wounds of their comrades, chased away flies, and even removed parasite eggs from them. They ministered to the worst cases for weeks. If the troop needed to move to a new location, they slowed their pace so that the wounded could keep up.

But sometimes, all their care was for naught and the wounded chimp died. Then its survivors adopted its orphaned children. It was usually the siblings of a deceased mother who did so, but sometimes even unrelated chimps would look after the little ones.[21] Still more surprising, there were even adoptive fathers. Normally, nothing is as foreign to a male chimpanzee as being a father. They have sex and don't spend another second being concerned about the consequences. In a fight with a different band, they sometimes even bite children to death. But here, male chimps are suddenly willing to care for their charges, support them in fights with others, and even occasionally share food with them. Boesch used genetic tests to show that in three out of four cases, the adoptive child was not related to its protector.

Caring for their wounded, and especially the rainforest adoptions, demonstrate how much chimps—who are usually only concerned with their own advantage—will do for others if the survival of the entire band depends on altruistic behavior. In the end, their devotion to their comrades benefits the caregivers as well, for every additional animal in the group is life insurance against a leopard attack.

The mutual caring among the apes of the Taï National Park is a striking example of how harsh living conditions can encourage the expansion of an innate inclination to be concerned for others. The chimpanzees don't require very much social intelligence to clean a comrade's wounds or adopt an orphaned child. They have no need to explore the inner life of another, because it's obvious what needs to be done. Nor is any extraordinary self-control required. Their ability to sympathize and their maternal instincts suffice.

But the animals make unusual use of these natural predispositions.

They not only empathize with the pain of others, they also try to assuage it. Males even suddenly display maternal behavior. Obviously, the constant threat from the leopards not only strengthens community spirit but also brings to light new aspects of every individual.

Humans behave the same way today. While in good years all of us are more preoccupied with seeing to our own affairs, we experience our interdependence when food becomes scarce or enemies attack. It is said that the reserved English have never been as friendly to one another as during the months of the Blitz. And those who experienced the end of the Second World War in German cities can recall the almost superhuman devotion of many of the so-called *Trümmerfrauen*, the "rubble women" who cleared the bombing damage brick by brick. After the shock of September 11, 2001, New Yorkers displayed a never-before-seen helpfulness and solidarity; there was even a relaxation of the usual tension between blacks and whites.[22]

The Move to the Savanna

Compared to the savanna, even a forest full of leopards seems a relatively cozy place to live. The conditions in open grasslands are much harsher, often brutally so. Nourishment is scarce, and competitors for it are all the more greedy. And nothing offers protection. Lions, jackals—and leopards, of course—can spy their prey from afar. A primate has almost no chance to escape. The climate, too, is hard on the savanna's inhabitants. Shade is rare, and for months, the ground dries up. Wildfires are frequent.

This was the environment in which our ancestors probably settled more than three million years ago. They did so because despite its hostility, the savanna offered certain advantages. Back then, members of the genus *Australopithecus* had long since mastered an upright stance. This forebear of ours had evolved into a runner while still living in the forest. And now he was better suited to life in the open grasslands than were other kinds of apes. At first, these protohumans could at least sometimes retreat to open forests or watercourses. But then, about two and a half million years ago, a climate change brought increasing drought to East Africa. Rivers dried up and forests turned into steppe. To survive, they had to adapt to the new conditions.[23]

Now the protohumans were more dependent on each other than ever before. In this environment, an individual by himself was lost. Only a sufficiently large group offered protection. Thus, the situation of our forebears resembled that of the Taï chimpanzees today, except with a much greater variety of challenges. All the more reason that their way of life—and they themselves—had to change.

Getting Through the Bottleneck

In the early stages of its development, humankind was repeatedly threatened with extinction. Analysis of deep-sea sediments as well as cores from the Greenland pack ice reveals that in the Neolithic and Mesolithic as well as in the preceding phase, climate often shifted dramatically.[24] At the end of the last interglacial period 115,000 years ago, for example, a period of drought suddenly began. The forests of central Europe burned and withered; where there had been spruces and beeches covering the land, sandstorms now blew across the steppe. Then the trees returned, soon to be replaced by an ice-age tundra.

In Africa, where at this time the first modern humans were struggling to survive, the climate swings were no less sudden. Again and again, they forced our forebears to give up their territory and venture out into new habitat. How devastating the famines and long marches must have been is written into the genes of all of us. So-called molecular archeology reveals that entire societies were extinguished in the turmoil of the Neolithic and Mesolithic. By comparing the genes of various humans, not only can one draw conclusions about how closely related they are, but also how numerous humans were in the distant past. The more contemporary humans differ from one another in genetic makeup, the more stone-age ancestors we collectively have. On the other hand, the more the genes of all contemporaries resemble one another, the fewer ancestors we can all trace ourselves back to. In fact, the genetic makeup of all contemporary inhabitants of the planet is astonishingly similar.[25] There are not even big differences between such widely separated peoples as the Eskimos and the Australian Aborigines. Consequently, during the longest period of its history, our species was very small in number,[26] seldom more than 10,000 according to the latest estimates by molecular archeologists.

Our forebears were not able to be fruitful and multiply until the population explosion of the so-called Upper Paleolithic Period, about 40,000 years ago, when new techniques for producing sharper and longer-lasting stone blades were developed. Until then, humans were simply extremely lucky. At least three times, their numbers must have been reduced to a dangerously low level. The first such "bottleneck" for the population curve occurred about two million years ago, when the genus superseded the protohuman *opithecus*. Another catastrophe was precipitated by the aforementioned climatic shifts in the Neolithic and Mesolithic. And the third widespread die-off was the terrible price the first modern humans paid for eventually subjugating the entire earth: Innumerable clans perished during the exodus from Africa about 70,000 years ago.

The only communities that were able to survive were those in which the members supported one another. Solidarity on all levels was rewarded. Families who stuck together better than others were able to produce more descendants. Clans that earned the trust of their neighbors could barter for what they needed. And small-scale societies in which individual clans subordinated their interests to those of the larger community as a whole wasted less strength on internecine quarrels.

But a troop in which individuals looked first to their own advantage was holding a bad hand in the chaotic prehistoric poker game. If critical numbers of members were killed because the others hadn't given them enough support, the group became too small to survive. It was just a question of time until the others died as well.

Community in the Nursery

We do not know the order in which evolution responded to these adversities. We do know that in the transition from *Australopithecus* to his successor *Homo rudolfensis*, brain volume increased by 50 percent. That was two and a half million years ago, when the first great climate change was happening in East Africa. At the same time, the first representatives of the genus *Homo* were beginning to eat more meat than their predecessors had. At first, carrion may have been an important source of nourishment. After all, men lack powerful fangs and claws, so hunting game would not have been very promising without tools. Our ancestors only overcame this handicap when they began to use weapons and hunt together. Thanks to their intelligence and above all their capacity to work together, *Homo sapiens* became big-game hunters.

It is still unclear what exactly it was that sparked this revolution. The American anthropologist Sarah Blaffer Hrdy hypothesizes that our ancestors' intelligence was developed through the raising of children by the entire community. Hrdy worked for years researching wild apes in the field and noticed an often overlooked difference between humans and most other primates. Among almost all the apes, the mother alone is responsible for her offspring, but in most human cultures, she assumes on the average only half the work of raising a child. The rest is shared by the father, grandparents, and other relatives, followed by friends, neighbors, and professional caregivers. From the first days of our lives on, many people work together to care for us.

In such shared devotion to the next generation Hrdy sees the source of our unique ability to cooperate with one another.[27] The advantages of communal upbringing are obvious. If a mother can temporarily leave her child with others, she can go hunting for food for herself and her offspring—an important advantage for survival on the bleak steppe. She can also more quickly recover from the rigors of giving birth and regain her fertility. The simple fact of communal child-rearing makes possible the luxury humans enjoy of a long, drawn-out childhood. Among the East African Hazda, for example, one of the last populations to still be using stone tools, the mother gives up her child as a newborn. Others then look after the baby 85 percent of the time. Anthropologists have found similar behavior among other traditional societies.

On the other hand, having multiple caregivers also caused changes in children, Hrdy conjectures. Whereas a baby ape has to make do with one mother, human babies see themselves surrounded by many faces from their very first day. Father and siblings, grandparents, aunts and uncles, and friends of the family—all of them want to not just admire the new arrival but care for it as well. At the same time, however, the relationship to all these grownups is more fragmentary than the bond between a young ape and the mother who gives it nourishment and warmth day and night.

There is only one way out of this dilemma for a human child: It must learn as soon as possible to assess and adapt to the moods and needs of others. That creates a new evolutionary pressure: The better a child understands her fellow humans, the greater are her chances for survival when nourishment and affection are in short supply. And so girls and boys who, by a mutation in their genotype, are endowed with a more pronounced social intelligence have an advantage. They are more likely to reach adulthood and reproduce than their socially inept comrades. And so the accidentally modified genes survive, and social intelligence steadily increases in the population.

Just as every healthy child learns to speak, all humans also have social talents inscribed in their genotype. How well the shift from one's own perspective to that of a stranger succeeds, of course, also depends on how much those talents are exercised during the first years of life. Studies by developmental psychologists show that children who grow up among older siblings or in a complicated web of various caregivers develop their social intelligence more quickly than an only child who spends most of its time in its mother's care. The anthropologist Hrdy sees this as a confirmation of her theory.

First Friendly, then Smart

Communal child-rearing, however, is only one path among many that could have led humans to their outstanding ability to cooperate. It is also conceivable that everything began with the communal hunt for large game such as lionesses engage in. But while the big cats work together only when necessary, over time early humans did so in more and more situations. Those who took part received their reward; at the same time, it became the custom to punish the shirkers.[28]

In one way or another, it was the cooperative individuals among our ancestors who were at an advantage and were able to reproduce in greater numbers. Perhaps all of these factors worked together to teach early humans to care for others.

Today, at any rate, it is innate for us to care about the welfare of others far beyond the bounds of simple helpfulness. Even small children behave differently than chimpanzees: As we have seen, when a chimp gets a banana, he doesn't care a bit if a fellow chimp gets nothing. If you give a child a jelly bean but give his playmate nothing, the child too will eat the candy. But already at two years of age, the child will feel better if they both get jelly beans. That was the result of experiments in which children were allowed to choose between the two possibilities.[29] If this inclination to fairness appears at such a young age, it is highly likely that it is innate.

This emotional constitution of our ancestors must have developed early—at any rate, long before *Homo sapiens* migrated out of Africa 70,000 years ago. For pronounced sympathy, the ability to empathize, and the willingness to share and work together are present in humans all over the world. After all, discoveries like the 100,000-year-old buffalo bones in the Klasies cave and even earlier evidence of communal hunts prove that even then, humans knew how to engage in complex activities together.

No matter whether it was communal child-rearing or hunting that led them to it, without a doubt our ancestors were socially intelligent first, before they could develop other intellectual capabilities. Until recently, however, many paleontologists argued the opposite: Abstract intelligence and language developed first, and only after that a sense of the needs of others. But this sequence seems unlikely when one considers how much energy our brain consumes. It uses almost a fourth of the nourishment we eat. Worse yet, it takes about fifteen years for young *Homo sapiens* to mature and be capable of procreation. The price of this very long childhood is enormous: Sons and daughters require costly nourishment, hinder their parents' freedom of movement, and are at the same time highly uncertain investments, since they are in constant danger of becoming the victim of starvation, illness, or an attacker.

So everything argues that our ancestors could only afford their large brains once they had learned how to minimize the risks of existence by working together. Before they were smarter than their cousins the chimpanzees, they had to improve their cooperation. Intelligence, language, culture—we owe all of these accomplishments to our sympathy and our ability to put ourselves in another's place. We humans became first the friendliest and then the most intelligent apes.

But now there was no turning back. The effort required to raise a human child had become so great that a mother could no longer master the task alone. And the energy demands of the brain had increased so much than the necessary nourishment could only be amassed if humans united their forces. Cooperation that went beyond the biological family became possible and necessary. Trust and friendship developed. The forms of association continually increased in size— from the clan to the larger community to the nation. And thus, besides individuals struggling to reproduce, a new factor came to determine evolution: the group.

CHAPTER 7

It's the Principle of the Thing

And the L*ORD* *said, If I find in Sodom fifty righteous within the city, then I will spare all the place for their sakes.*

GENESIS 18:26

AN ANONYMOUS BENEFACTOR OFFERS YOU $1,000, BUT with two conditions: You have to give me some of it, and I have to accept the gift. If I refuse, you get nothing. How much will you offer me?

Half? You are a generous person. But we don't know each other and will probably never see each other again after the transaction. Maybe you don't have to give me that much. Surely I would be satisfied with $400, wouldn't I? Or maybe even $200? Or $20? Then you could keep $980 for yourself. After all, as long as I play along, it doesn't really matter to you if I think you're stingy—knowing, as I do, how much you're keeping for yourself.

If you were in my place, would you agree to such an offer? Wouldn't you be annoyed? Because why should one person should get so much more than the other? Is your partner going to buy himself a cashmere overcoat and fob you off with a lousy T-shirt? You don't want to be played for a patsy, so you refuse.

If you are the giver and want to make a reasonable calculation, you'll have to ask yourself where my acceptance limit lies. If you assume that I will act purely logically, like *Homo economicus*, then the answer is obvious: One dollar is enough for me. Because if I refuse your offer, neither of us will get anything and one dollar is better than no dollar at all. So if I say no, I'm hurting myself simply to punish you. But you know very well that I'm no *Homo economicus*, and neither are you. It would probably never occur to you to make me such a shabby offer.

And that would be completely normal behavior. A thousand people have played this game and most of them offered between $400 and $500. And they do well to do so, for half the receivers turn down offers under $300. If higher amounts are involved, subjects in the experiment sometimes even refuse offers equal to three months' salary if they think the division is unfair.[1]

The game is called Ultimatum and was invented in 1981 by the economist Werner Güth, currently doing research at the Max Planck Institute of Economics in Jena, Germany. It measures how keen our sense of justice is, or put differently, how much we're willing to sacrifice to punish someone who has violated our feeling for what is fair. That's exactly what a receiver does who turns down an offer he considers shabby. We want justice—even if we have to bleed for it ourselves.

Justice and Altruism

Philosophers understand justice to be an appropriate balance of interests that anyone can demand. In general, we praise as selfless men and women who give to the needy. But there is much less unanimity about whether people who demand justice are altruists, which is astonishing, for justice places a greater demand on our morality than does charity. One can freely decide whether to be generous or not. We can give something today, nothing tomorrow, and something again the next day. But where justice is concerned, the norm is that one *has* to share. And if anyone refuses, she must be punished or the norm will soon be worthless. The demand for justice has an effect on a community similar to outside pressure: It makes it harder for egocentrics to refuse to cooperate and thus stabilizes social cohesion.

But justice has costs. The person who gives something away is not the only one who pays a price. The person who punishes the uncooperative also sacrifices something. In real life, as in the game Ultimatum, there are costs and trouble associated with imposing sanctions. In addition, one incurs the hostility of those who get punished. Thus the struggle for justice always demands altruism.

Economists will object that people only call for justice in order to put themselves in a better position. If we sometimes forego a short-term advantage for the sake of fairness, they will say, it is only because it will pay off in the long run. The other person presumably realizes that he can't treat us that way and will make us a better offer next time.[2]

But neither the behavior of people playing Ultimatum (where the players will never see each other again) nor of people in everyday life substantiates this theory. Employers know that it usually pays to offer workers a fair salary.[3] Like the giver in Ultimatum, their business is dependent on the cooperation of the other participants—in this case, their employees. And lack of cooperation is something that the employers cannot sue them for. Nothing and no one can force employees to do more than what the job requires. Only those who feel they are being fairly treated will work more than absolutely necessary—and they do it because they feel an obligation. Studies have shown that more pay—even for a job that lasts only one day—leads to more output. On the other hand, it's a bad idea to cut wages during bad economic times. Personnel immediately reduce their effort, even though by so doing they act against their own best interest, just like the receiver in Ultimatum who refuses to accept a low offer. Employees only accept such reductions when the company is on the brink of bankruptcy.

There Must Be Punishment

A single day spent in a small-claims courtroom is enough to dispel once and for all the argument that we only seek justice when it benefits us to do so. Like Michael Kohlhaas in Heinrich von Kleist's story,[4] plaintiffs demand that defendants who have done them wrong be punished, without regard to the plaintiffs' own losses. Customers sue a retailer who they claim sold them defective goods valued at less than $65. A retiree demands compensation because during a group tour, the promised air-conditioning on the tour bus malfunctioned. And a handyman sues a client for two hours of work he performed but hasn't been paid for yet. The lawyers and the court always earn many times more than what a successful plaintiff can expect to recover in the best of cases. And since the plaintiffs and the defendants regard each other as crooks, they will certainly never do business again. As unsympathetic as one may find the litigants, their behavior certainly cannot be described as motivated by the hope of personal advantage. It's a matter of principle.

Not even corporate lawyers, by the way, behave like *Homo economic-us*. That was the conclusion of the social psychologist E. Allan Lind after he evaluated a whole mountain of American court documents. Although in these cases, plaintiffs and defendants were not individuals squabbling over small claims, but large companies with seven-figure sums hanging in the balance,[5] they obviously allowed themselves to be guided by their sense of justice. Whether the parties before the court accepted a settlement depended not on how advantageous the bottom line was for them, but rather on whether they thought the proceedings and the settlement were fair. If they thought they were being taken advantage of, they would continue to litigate even at the risk of ending up with horrendous court costs.

And it's not just in court that people accept sacrifices to achieve justice. People perform impressive acts of civil courage every day when, for example, they intervene to stop discrimination against people of a different color. Or recall the story of Rosa Parks, an icon of the American civil rights movement, who preferred to go to prison rather than surrender her bus seat to a white passenger, which was the law in Montgomery, Alabama. During the following year of 1956, more than 40,000 blacks refused to ride city busses. Instead, they walked long distances or spent their money for taxis until the Supreme Court ruled that Rosa Parks had been right and the city officials had to give in.

In our own day, the indomitable Burmese oppositional figure Aung San Suu Kyi proved that the longing for justice could be even stronger than the ties to one's loved ones. Except for short interruptions, the Nobel laureate spent over fifteen years under house arrest while the Myanmar dictators refused entry to her two sons. Her British husband Michael Aris saw her last at Christmastime in 1995, shortly before he was diagnosed with incurable cancer. Despite all the diplomatic efforts of even the General Secretary of the United Nations and the pope, the generals in Rangoon refused to grant the dying man a visa to visit his wife again. Instead, they pressured Suu Kyi to leave the country. Since they would surely not have allowed her to return to Myanmar, she declined the chance to say good-bye to her husband; the struggle of her people for freedom took precedence. Aris, who married Suu Kyi in 1972, died in 1999. In the last ten years of his life he was allowed to meet with his wife only five times.

No political activist ever offered his followers happiness, the Swiss author Max Frisch noted in surprise. It was the hunger for justice that drove people into the streets.

The Nicest People in the World

But what do we consider just? Everyone can decide for themselves what will make them happy, but we must reach agreement about what is just. And it's always going to be about sharing. Whether we're talking about seats on a bus, the power of the state, or dividing up a pie, the key to how each person gets his share is not obvious. Every act of apportionment functions only as long as no one challenges it. What we find just, therefore, is nothing but a question of the norm, the agreement that has become established in a society.

But who or what determines the norm? Researchers all over the world have addressed this question using the Ultimatum game as their tool. It allows them to question people about their conception of fairness in a simple and easily comparable way. Moreover, the responses are naturally honest, because if fairness is important to players, they have to pay more for it in the game. Thus the game also makes apparent under what circumstances unselfishness flourishes or greed predominates.

For their comparative study of egocentrism and altruism in various nations—the largest such study to date—the anthropologist Joseph Henrich and his colleagues traveled to the most distant corners of the world.[6] They gathered data from college students in Los Angeles, Tokyo, Jerusalem, and on the Indonesian island of Java. They questioned East African farmers and Mongolian sheep herders, visited indigenous communities in the Chilean Andes and along the Amazon in Peru. Their only failure was in a village in the forests of Papua New Guinea, where a woman declared the experiment the "work of the devil" and pulled a knife on the intruders.

Wherever they were, they had their subjects participate in a game of Ultimatum in which the amount at stake was always equal to two days' pay. And everywhere, people proved to be remarkably generous. In industrially developed countries as well as in the urban areas of the developing world, the games always ended with the same result: On average, the first player gave away half the money while the second player rejected offers of anything less than one-fifth of the original sum. This result is astonishing in and of itself. However much the city dwellers differed in their language, religion, or everyday customs, their idea of fairness was remarkably uniform.

In traditional societies, however, the amount that both players considered fair varied enormously. According to Henrich's study, the most generous people on earth are the Lamalera, Indonesian whale hunters who regularly offered their partners in the game almost two-thirds of the money. The least generous were the Machiguenga, an ethnic group living in the Peruvian rainforests. They kept three-fourths of the total for themselves, which their partners considered completely acceptable. But nowhere did people behave even close to the way traditional economic theory would have predicted. Even the Machiguenga are still a long way from being pure egocentrics, so deeply rooted is the social instinct in all humans.

In experiments, people almost always make the same decisions they would make in real life. Behavioral scientists designate the Au and the Gnau, two ethnic groups in Papua New Guinea, as "hyperfair," just like the Lamalera. That means that they often offer their partner in the game more than half of what they have. But unlike among the Lamalera, here the high-minded donors often meet with resistance. Among the Au and Gnau, it is a status symbol to give large gifts, but accepting them means abasing yourself. The Ache in Peru, on the other hand, almost always offered half and never turned down an offer. These hunters are used to sharing their catch among all the families in the group, and no one can afford to turn down the daily ration.

Trade Makes for Generosity

What makes the Lamalera so generous, and why are the Machiguenga so stingy? To explain the differences between societies, Henrich and his colleagues controlled all possible factors, from the population size of the villages, to their social structure, to their ability to keep a secret. None of those factors accounted for the difference. It was only when the researchers compared the extent to which people in each culture carried on trade and cooperation with people outside the individual clan that they found the answer.

The Machiguenga are individualists. Each family lives separately, ignorant of the joys and sorrows of other families. One of the infrequent occasions when members of different clans meet is to fish. Together the Indians dam up the rivers and poison the water. But as soon as the dead fish float to the surface, everyone dashes in and tries to grab as many as they can. This quite asocial behavior is reminiscent of the way chimpanzees hunt the colobus, but it doesn't bother the Machiguenga.

The Lamalera, on the other hand, would starve if they were similarly selfish. The barren volcanic rock of their homeland offers hardly any arable land, which makes them dependent on whale hunting, an endeavor that requires everyone's participation from beginning to end. Their boats, powered by oars and palm-leaf sails, set out as a flotilla, each one with a crew of at least ten. The captain of each cannot always ensure that all crew members are from the same clan, since he needs experienced seamen. Only when the steersman, the harpooner, and the oarsmen work precisely in concert do they have a hope of catching a whale. Often enough, one of the boats capsizes during the hunt. Then the others come to its aid. Once the prey has been landed, its flesh is not just divided among the hunters; the boat- and sail-makers also receive their portions according to elaborately worked-out rules. If anyone violates them, he gets excluded from the hunt for a certain period and must go hungry.

Obviously it is variations in the ways of life that make humans in tribal societies generous or not. The Machiguengas' lifestyle is very primitive. In the Peruvian forest, every family must feed itself, and so people are not used to bargaining. They have no experience of how to compare the value of different things, nor any notion that cooperation can benefit everyone. People obviously need such experience to hone their sense of fairness.

The Lamalera, however, are generous because, from a very young age, they learn how much they need one another. They also trade with people outside their community and sell their wares at the market. In general, Henrich's comparisons showed that people are more willing to share the more they depend on others outside their family, either to do business or because their survival depends on cooperation in the group.

Practice Makes Fairness

Even in the highly developed societies of modern cities, we are so used to trading and making deals that we cannot get along without a strong standard of fairness. Thus our definition of fairness is determined less by the intellectual sphere of our culture than by the way it does business. People who live in large cities the world over earn their living in similar ways. Be it in Frankfurt or Jakarta, Chicago or Mumbai, most people sell their labor and use their wages to purchase goods in shops. Thus the same concept of fairness became established in the cities of Europe and America, on the island of Java, and in the Middle East.

Of course, the standard only determines how *most* people behave. As always, there are deviations at the top and the bottom of the scale, i.e., especially altruistic and especially selfish people. These differences are often connected to the life circumstances of such individuals.

Studies have shown that students of economics are more inclined to be selfish than the average person.[7] When these future managers and bankers play Ultimatum, they are both stingier when they make offers and more willing to accept offers that others indignantly reject. In other experiments, they evinced few inhibitions against cheating a partner for personal advantage. And in contrast to other college students, economics majors rarely thought of making charitable contributions.

Of course, this behavior could be explained by assuming that a lot of naturally greedy people want to be business experts, but there is evidence that this is not the case. In general, college students become more altruistic in the course of their education. Apparently they find out that it's a good idea to do things for others. It's only economics majors who show no sign of a keener sense of fairness or an increased willingness to cooperate. What does increase during their studies is their cynicism. This was the finding after they were asked questions such as, "Would you return a lost wallet to its owner?" The students seem to be reacting to the image of humanity presented to them in their courses. (After the first such study was published in 1993, the reaction at some colleges was so outraged that the study's principal investigator, the American economist Robert H. Frank, found it necessary to explain that he was *not* asserting that majoring in economics made you into a potential serial killer—it just influenced your behavior in certain situations of social conflict.)

Obviously, our sense of fairness is like our leg muscles: Both are strengthened by training. A study of Swiss bicycle couriers came to that conclusion. Some couriers were paid by the hour while others were paid according to how many deliveries they made. When tested, the couriers who received an hourly wage proved to be much more altruistic, while those who did piecework were much less generous. They had obviously become used to everyone looking out for number one.[8]

We Are Opportunists

One can argue about the essence of fairness, but all points of view boil down to one fact: It's obviously unfair when one person gets something for free that everyone else has to pay for. That's why people who sneak onto public transportation without paying for it always get such dirty looks when they get caught. For the same reason, most people are disgusted when CEOs run their companies into the ground and then walk away with golden parachutes. Because, of course, the money has to come from the pockets of their customers and employees—or from taxpayers. In other words, people find it unfair to be cheated by others.

And that is exactly what makes cooperation in a group so difficult. With the help of another experiment called the Free Rider Game, one can investigate how trust in others and a sense of fairness are translated into solidarity. Let's say that a group consists of six participants, all of whom contribute money voluntarily and anonymously to a common pot. After each round, the investigator triples the amount collected and divides it equally among the players. If they all contribute $10, each gets $30 back. So it's worth it to put money in the pot—but even more so not to put any in and just collect after each round.

The contribution each player makes to the common pot can stand for the price of a subway ticket or the productivity of an employee. The payback after each round stands for the value added in each case: The passengers get from here to there; the company makes a profit and can pay its employees. But since the subway follows its schedule in any case and union contracts require the company to deposit employees' earnings in their bank accounts every month, there's a temptation to weasel out of paying one's fair share. It works just the same way in the Free Rider Game: If only five players pay $10 in instead of all six, after the investigator has tripled the pot it contains only $150 instead of $180. Each player gets $25 instead of $30. Everyone is still ahead of the game, but the free rider has made out the best. He now has $35 in his account (the $10 he held back plus the $25 he received).

The paradox of the Free Rider Game is that although the group as a whole earns more if everybody pays in, from the perspective of the individual it's always more advantageous to hold onto one's money and also collect an equal share of the common pot—to the detriment of the group. In our real-life example from public transportation, the subway

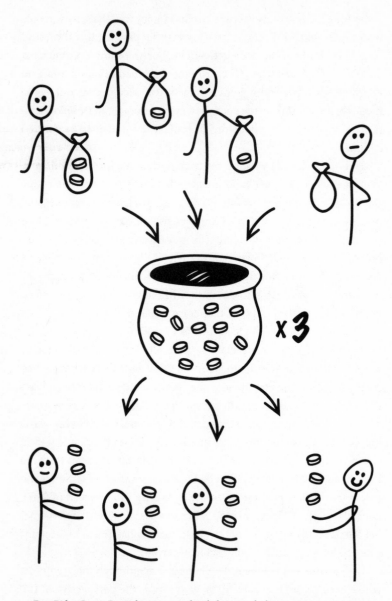

Free Rider Game. Four players get to decide how much they want to pay into a common pot. The sum of the contributions is tripled and divided evenly among all players. If a player like the one on the right contributes nothing, he cashes in nevertheless. Although his strategy is the most profitable, most people contribute—as long as others do, too.

company earns less money if a lot of people are jumping the turnstiles. It has to hire ticket checkers and raise prices to do so.

Happily—and despite the temptation—not even one-third of all people act with consistent selfishness. The majority (more than half of all subjects in the experiment), on the other hand, are opportunists. They would like to do something for the common good, but not at any price. At the start of the game, they behave generously as long as others are also being cooperative. But they start to get stingy if their fellow players pay in less than expected. That is precisely our norm for fairness: We're quite ready to do something for others and for the community as a whole, but we're unwilling to be exploited.

The remaining fifth of humanity consists of extremely altruistic men and women. They sacrifice themselves for the group even when they are surrounded by slackers. But most people simply mirror others' behavior. If others do their share, we do ours. But if others act selfishly, we're going to look out for our own advantage. We're less a *Homo economicus* than a *Homo reciprocans*, humans driven by the wish to reciprocate. Most of us practice selflessness with a proviso; we're provisional altruists.

Shape Up or Suffer the Consequences

As so often happens in real life, people in the Free Rider Game experience disappointment: Some egocentrics can't wait to exploit the goodwill of the majority. And since their behavior violates the norm of fairness, cooperative participation in the game soon begins to fall apart. When one person repeatedly takes without having paid in, the others also refuse to contribute. Cooperation usually ceases after only a few rounds. Now all the players are only looking to their own advantage and hanging onto their money.

But the dynamic is profoundly altered if they have the chance to punish the free rider.[9] Their revenge, however, costs money. For example, a player must pay a dollar to have a genuine or suspected cheater fined three dollars. Nevertheless, players have frequent recourse to this punishment. And it works: Now they all contribute their fair share. The contributions even increase from one round to the next. Suddenly, the norm of fairness has traction. It prevents the greedy from plundering the generous and so creates the preconditions for cooperation and sharing. This is one of the most important things that distinguishes us from the chimpanzees, who lack any such regulative norm.

Whoever violates the norm must pay a penalty. And so players with goodwill attack the greedy at their weak point: The penalty makes it expensive to violate the rules. In order to avoid that expense, it's better to play fair. Justice is the altruists' strongest weapon.

Still, it is astonishing that the mechanism works. After all, the norm doesn't solve the problem of freeloaders. As in the case of the prehistoric buffalo hunters, it simply raises it to a higher level. For it would never occur to anyone who thinks logically only of his own benefit to play the role of policeman. Because when a freeloader is punished, it's only the enforcer who pays the price for doing so, although everyone enjoys the benefit. Thus it would make more sense to leave the dirty work to others. That's why self-interest cannot be the reason why people punish miscreants in order to maintain group morale. Enforcers are plainly altruists.

The Counterweight to Greed

What internal mechanism is at work when we punish someone? The Swiss neuroscientist Dominique de Quervain and his colleagues found an unflattering answer to this question.[10] In an experiment, they had their subjects play the trust game. Whenever a player abused a partner's trust, the latter was permitted to punish her. The punisher had a small sum of money deducted from his account, the miscreant a larger amount. The investigators recorded what went on in the brain of the punisher during the transaction. The reward system was activated, the cerebral circuitry already described in Chapter 3 that causes sensations of pleasure at the prospect of something good—food, for example, or sex. Revenge is literally sweet. The chance to really give a cheater what's coming to her makes us feel so good that we're happy to part with a few dollars for the pleasure. The stronger the signal from the reward system, the more de Quervain's subjects were willing to pay for the chance to punish. And as we saw in Chapter 4, men—but not women—seem to enjoy it when the cheater gets a painful shock as well.

Several experiments revealed the step-by-step activation of the will to punish. The first step is resentment at another player's untrustworthy behavior. The front cortex of the cerebrum becomes active.[11] This region is responsible not only for anger but also for one of the strongest feelings of all: disgust. When we must swallow bitter medicine or a cockroach runs over our hand, revulsion is plainly written on our face. And we curl our upper lip in exactly the same way when someone is not playing fair.[12] At such moments, the brain even sends the digestive system the signal to vomit, which is, however, suppressed as the mechanism of punishing proceeds. Even if the nausea usually remains unconscious, it is more than a colloquial metaphor when we find some people "disgusting."

Disgust makes humans lose sight of all their other aims as well as their manners. Since it is part of the impulse to punish, it is not surprising how implacable avengers often are. The more active their cortex is, the more the subjects in the experiment invested to punish a miscreant.

Our negative emotions are not, as one might suppose, released by our fear of being put at a disadvantage by someone's unfair behavior. Rather, the catalyst is someone else's violation of the norm, for the reaction happens even when we only witness someone breaking the rules to the disadvantage of a third party. For example, we are disgusted to learn that a confidence man has robbed an old lady. Conversely, if subjects in an experiment know that it is only a computer that is being unfair, anger and disgust do not enter the picture. There would be no point in trying to discipline a machine.

In the second step of the activation process, the brain's wish for revenge becomes a deed. Responsible for this step is the so-called dorsolateral prefrontal cortex. This region of the cerebral cortex, located behind the right side of the forehead, suppresses the striving for one's own advantage in the interest of justice. When this mechanism is not functioning, pure egocentrism takes over. But the unselfish desire for fairness is so strong that a human brain must be massively disturbed to make it forget that desire. Daria Knoch, a neurologist at the University of Zurich, was able to achieve that result by radiating the right frontal lobe of her subjects with strong magnetic fields. Suddenly, her Ultimatum players were willing to accept shabby offers.[13] They were well aware of how unfair it was, but their desire to realize even a small profit was now irresistible. Only when the magnetic fields were turned off did their preference for fairness return.

So do unscrupulous egocentrics have a brain defect? In some cases, they probably do. In fact, the behavior of Knoch's subjects was similar to that of patients whose dorsolateral prefrontal cortex has been injured in an accident or by a stroke.[14] These patients are quite conscious of violating a norm when they obsessively lie, cheat, and steal. But they are helpless against those impulses. Their greed has no counterweight.

Tempting Punishments

To be sure, it remains an open question *why* our brain is structured to react to unfairness with disgust and feel schadenfreude at a just punishment. It's easy to understand that we protect ourselves against exploitation by cooperating only as long as others do, too. But it's harder to explain why we punish others even at our own expense. Animosity and revenge must offer advantages, or they would not have survived natural selection.

The deeper reason for why punishment can survive within a group is that its cost-benefit ratio is so advantageous: It costs less to hold others to the standard of fair play through occasional sanctions than to reward them for cooperating. For example, it would always be cheaper to leave your car in a no-standing zone than in a parking garage. You could save something like thirty-five dollars per day by doing so. But the disadvantage is that if everybody parked where they wanted to, the traffic flow would fall apart. Society has two ways to avoid chaos: Either it can increase the attractiveness of parking in the right place by building free parking garages, or it can pay people to issue tickets to illegally parked cars. Of course, the second way is by far the cheapest. The cost to each individual in the society is minimal, and a single meter maid can keep hundreds of drivers in check and the traffic flowing more or less freely in an entire neighborhood.

Since punishments are a deterrent, for the majority of us provisional altruists they are a powerful lever to maintain the norm in the group. This is exactly the effect that was shown in the Free Rider Game experiments described above. In the first round, there are severe punishments; after that, participants obey the rules of their own accord. Only occasionally does a player need reminding that the norm must still be taken seriously.[15] Thus as soon as a deterrent proves its effectiveness, incorporating punishment is a very economical way of behaving altruistically. It's true that someone who administers the punishment for every violation will always be in a worse position than someone basically willing to overlook a wrong. But since the difference is small, this kind of altruism is able to survive, even in large groups that otherwise would immediately be plundered by freeloaders.

That's why it's only surprising at first glance that people voluntarily choose communities in which there is punishment.[16] In free rider experiments conducted by the Erfurt economists Bettina Rockenbach and Özgür Gürek, participants were allowed to decide anew in each round which of two groups they wanted to belong to. In one group it was permitted to punish miscreants, but not in the other one. At first more participants joined the unconstrained group—they weren't masochists, after all. But as soon as word got around how much more everyone earned when they all conformed to the norm, the threat of punishment developed a phenomenal attraction. After thirty rounds, the group without sanctions was completely depopulated. Even the last free spirits had buried their anarchic ideals and switched to the camp of the guardians of morality.

Does this mean we must sacrifice our freedom to obtain justice, profit, and space to walk on our sidewalks? No; in fact, quite the opposite. Punishments can only preserve the morale of a group when its members submit to them of their own free will.[17] Everyone must be able to choose to leave the strict community; that's how the well-intentioned and the freeloaders are automatically separated. Those who remain very likely intend to cooperate. If morality nevertheless crumbles over time because too many group members are not willing to act as enforcers, the just individuals will leave. Where punishments help, they are paradoxically hardly necessary.

Of course, that assumes that various communities, including the possibility of not joining a community at all, are all in competition with one another.[18] Things get more complicated when frustrated members are not able to leave. Then there is a danger that the freeloaders can hold everyone else in the group hostage, because whoever does not act as an enforcer is always at an advantage. Players who insist on enforcing the rules can no longer threaten to leave if or when morality breaks down. And then the tragedy of people of goodwill unfolds: Even participants who would actually like to play fair must violate the norm unless they want to be at a hopeless disadvantage.

Cascading Gratitude

Since most of us are provisional altruists, how selflessly we act depends on the circumstances. And often it is trivial things that determine whether group members work together or all follow their separate paths.

For instance, whether citizens pay their taxes or not depends less on how much of their income a government demands than on whether it has let them decide ahead of time what a fair tax rate is and how the money is to be spent. That is the conclusion reached by the Zurich economist Bruno S. Frey in a comparative study of taxpayer honesty in the various cantons of Switzerland.[19] Each canton is organized like a mini-state, and politics vary from one to the next. Inhabitants of Geneva have rights similar to those in Germany: The most they can do is vote for their representative every few years. In other cantons such as Basel-Land, citizens have more decision-making responsibility. Every few months, they have to vote on education budgets, new highway construction, or tax laws. If the cantonal government wants to spend a considerable amount, they must go to the people first. In such cantons, tax inspectors have much less to do. Even when tax rates are just as high as in other areas, almost all the citizens of Basel-Land pay them willingly. Obviously, the control they exercise over communal affairs reinforces their sense of fairness. They do not fear that they're being exploited by the state and so are more willing to do their share.

What we think of our fellow citizens also has a big influence on how altruistic we are. If our neighbors' house burns down through no fault of their own, we're happy to take them in. But if the same neighbors are out on the street because the father has gambled away all their money and can't pay the mortgage, we're not as eager to help. Yet the need of the neighbors and the effort needed to help them is almost the same in both cases.

But our image of humankind has much wider implications. Whether it leads us toward generosity or stinginess, our behavior elicits similar behavior in others. In the Free Rider Game, all it takes is for one participant to have doubts about the honest intentions of the others—whether justified or not—and stop paying in. Soon other players, well-intentioned up to then, will follow suit. On the same principle, corruption and tax cheating can spread in a society like infectious diseases. If somebody thinks that others are committing the same "trivial offense," his wish not to put himself at a disadvantage will triumph over his loyalty to the group. And so he gives others a reason to act selfishly as well. The morality of a group is a question of trust; without that, cooperation collapses. Participants fall one after another, like dominoes.

Fortunately, however, our image of humankind can also encourage cooperation. If *Homo reciprocans*, attentive to the mutuality of trust, considers others to be honest, then he will be, too. If you would like people to be altruistic, you want to give them the feeling that they are surrounded by altruists. Clever pastors have long known that it helps to prime the collection plate with some cash before passing it around. Their experience has been confirmed by an experiment with several thousand university students in Zurich.[20] They were asked whether, in addition to the quite hefty fees they are already charged each semester, they would be willing to make a voluntary contribution to a solidarity fund for foreign students studying in Zurich. Although most students have to work part-time while they're attending university, their response was overwhelmingly positive. Almost two-thirds of them contributed. And that percentage rose when the economists conducting the experiment added to some questionnaires the statement that most students were contributing. Another group was falsely informed that only a minority of students were making contributions, and the contributions of that group promptly diminished.

An organ donor in Michigan set off an almost unbelievable cascade of altruism in 2007.[21] The man, who chooses to remain anonymous, was twenty-eight at the time and decided to donate a kidney to anyone who needed it, without expecting anything in return. Every year, about a dozen altruists in the United States submit themselves to this operation—which although not usually dangerous does involve a painful recovery period—to benefit utter strangers. (In most European countries, including Germany, voluntary organ donations are legal only between members of the same family.)

The donated kidney was transplanted into a fifty-three-year-old woman in Arizona. Her husband, unable to give her a kidney because of incompatibility, thereupon offered to donate his kidney anonymously to someone else who needed one. The recipient was a young woman in Ohio, whose mother then donated an organ. And so the chain continued, crisscrossing America (and its racial boundaries). Brothers donated because someone had helped their sisters, friends submitted to the operation because an anonymous donor had saved their buddy's life. No one was forced to undergo an organ removal, and donors often waited months for doctors to find a compatible recipient to whom they were *permitted* to donate a kidney. After a total of twenty-two operations, the chain ended—at least for the present—with an organ donated by a young African-American woman in Ohio. Not a single donor gained any personal advantage from the donation. They were all simply moved by their gratitude that some unknown person had given their loved one such a tremendous gift.

How Rewards Undermine Morality

This wave of generosity would probably never have occurred if money had been involved. For pecuniary rewards undermine people's will to do something for others, whether voluntarily or from a sense of duty.[22] The English sociologist Richard Titmuss was already alerted to this phenomenon in the 1960s, when some of the health services in his country began to pay blood donors. The change in policy had the opposite of the desired effect: Donor numbers went down.[23]

Parents have similar experiences when they begin to pay their children for household chores they have up to then been expected to do without remuneration. Once your daughter can increase her allowance by mowing the lawn, she won't touch another garden tool without some reward. And she'll probably do a messier job of mowing than when she was simply expected to do it without pay.

Youngsters in Israel go from door to door once each year to collect donations for charities. They get no other reward than praise. When Uri Gneezy, an economist doing research in Haifa, promised them a small percentage of their total as encouragement, their take fell by more than one-third.[24] For they had gone from being altruistic volunteers to badly-paid workers. When Gneezy raised their percentage to 10 percent, the results improved a bit, but they were still far below what the volunteers were bringing in. For a selfless deed, we expect recognition. After all, we're proud of ourselves. Compared to that, a bit of money is a weak motivator. Gneezy's conclusion: "Pay enough or nothing at all."

Money carries a message. It changes the norm of a community. Where there is a reward, people have to assume that other members of the group will not work for free. And that's when voluntary cooperation breaks down, to be replaced by more or less advantageous negotiations. It is difficult to back out of this revaluation. Our desire for fairness rules out accepting nothing when others are raking it in. Moreover, the status of work is transformed. From an activity we pursue of our own free will, it turns into a duty. Why would somebody pay us for something we're happy to do voluntarily?

If one treats people as egocentrics, they'll become egocentrics. This is illustrated by an absurd anecdote about a clever Jewish tailor whom religious fanatics were trying to drive out of town. Every morning they would gather before his shop to scream obscenities. When the situation threatened to get out of hand, the tailor had a bright idea. He emerged from his shop and handed each of the mob a dollar "as a bit of compensation for your trouble." The next day, they reappeared in front of his shop and demanded their money. "Sorry," said the tailor, "I can't afford to give you a dollar today, only a quarter." The disappointed rowdies took the coin and started their usual yelling. When they showed up again the next day the tailor said he only had a penny for each of them. The leader of the mob was furious. "We're not about to scream ourselves hoarse for a pittance!" They disappeared and never returned.

CHAPTER 8

Us Against Them

Neither was there any among them that lacked: for as many as were possessors of lands or houses sold them, and brought the prices of the things that were sold, and laid them down at the apostles' feet: and distribution was made unto every man according as he had need.

ACTS 4: 34–35

GEORGE PRICE WAS ONE OF THE MOST original thinkers of the twentieth century and also one of the most enigmatic. The American chemist seemed not to be made for the ordinary, unexciting life of a researcher. Early in his career he worked for the Manhattan Project to build the first atomic bomb, and then he shuttled back and forth between universities and the private sector. At the same time, he made a name for himself as an unconventional science journalist. In 1955, when computers were still printing out endless series of ones and zeros, he was already predicting the advent of the computer mouse and virtual images on a screen. In passionate articles, he argued that the US government should give every Russian two pairs of good shoes as a reward for the Soviets' withdrawal from occupied Hungary and fulminated against the adepts of extrasensory perception. Price was such a militant atheist that it destroyed his marriage to a devout Catholic. When he received a generous compensation for pain and suffering after an unsuccessful operation for thyroid cancer, he boarded the *Queen Elizabeth* in 1967 and set off for London to begin a new life.

There he holed up in libraries and began to mull over the question Darwin had already raised: Did evolutionary theory leave room for altruism? He happened upon two publications in which the geneticist William D. Hamilton had been at pains to show that altruism was only an advantage among blood relatives.[1]

Price checked the figures and discovered all by himself what Hamilton had overlooked in his mass of calculations: If altruism can flourish among relatives, it can do so in any random group. Price devised a new equation.[2] His formula was simple and beautiful—and so comprehensive that it could be applied not just to biology but to any competitive situation. Price boasted that it could explain why we prefer one decision to another as well as the rise and fall of the Roman Empire.[3]

Competition Engenders Altruism

He wasn't exaggerating. Science is indebted to Price for a decisive push toward understanding how human societies work and how communal life has shaped our being.

Darwin had already recognized that the question of altruism's role in evolution could be divided into two parts: First, how effective are altruists at passing on their genes *within* a group? And second, what are the chances of survival for groups with more altruists than other groups? But the father of evolutionary theory concentrated only on the first question and reached the unsatisfactory conclusion that altruistic humans would not be successful at propagating their genetic material because they were at a disadvantage in relation to selfish members of the same group.

But Price was able to prove mathematically that Darwin's conclusion was wrong. He showed that the disadvantage that selfless individuals in the group have to accept is more than compensated for by the advantage they bring to the group *as a whole*. Thus groups in which altruism is prominent are able to prevail over groups in which selfishness reigns, and because that is so, altruism is able to spread. Of course, there is one condition: Competition *between* groups must be stronger than competition *within* a group. In that case, the advantage on the group level outweighs the differences between individuals. In other words, natural selection takes place on several levels simultaneously.

An example is the competition between two groups described in the previous chapter, in which one group allows its members to do as they please, while the other punishes those who violate the norm. Although superficially it would seem advantageous for any individual to belong to the first group, more and more players soon abandon it to join the second because it holds together better.

Every basketball league illustrates the principle that Price discovered. For the career of a given player, the most important thing is how successful his team is; where he stands in the pecking order of his team is less important. A team of egomaniacs will never win the championship, not even if all its players are as talented as Michael Jordan. It's true that every team needs some stars—at least, it does as long as those stars fulfill the hopes placed in them. But to do that, they are dependent on their backcourt and bench, the players who aren't in the spotlight but still contribute to and profit from the team's success. As soon as too many of them also try to play first fiddle, the system breaks down; the team loses against other teams that play together better, and the market value of even a Michael Jordan can sink. Now the more selfless players of the team with less individual talent have their chance.

The example of a basketball team, however, leads us to another condition that must be fulfilled before altruism can triumph. It's not enough that a population breaks down into separate groups competing against one another like sports teams. These groups must also be variously constituted. Because if everyone everywhere behaved cooperatively, competition on the level of groups would become inoperable: All teams would play equally well. Now everything again depends on one's advantage within the group and the egocentrics gain the upper hand. George Price showed that as well.

The Death of an Altruist

The famous geneticist William D. Hamilton, whose article had inspired Price, was immediately convinced by the latter's findings. He changed his mind completely, now convinced that selflessness was an evolutionary advantage even outside one's blood relations. The scientific outsider Price won high praise from other experts as well. He received state subsidies and was given an office in the prestigious genetics laboratory of University College London.

But now Price's life took another unexpected turn. In the summer of 1970, he reported having had a religious experience, explaining to friends that one June day he simply found God. The proof of God's existence was to be found precisely in his former atheism—and in mathematics. He argued that the probability of his conversion was "astronomically low."[4] Thus a higher power must have been involved. He spent the following year working on an interpretation of the four Gospels and a new chronology of Passion Week.

In addition, Price became absorbed in game theory. In another spectacular article, he laid the groundwork for many of the ideas treated in Part I of this book.[5] But by the fall of 1972, while still working on that essay, he had already begun to turn his back on science. Now, devoting himself to the needy seemed to make more sense than any of his research, even in theology. He took homeless people into his house and made a bed for himself in his laboratory. He even followed the Sermon on the Mount literally, no longer taking thought for the morrow, and gave away his coat and his watch. "I am now down to exactly fifteen pence," he wrote to John Maynard Smith, the co-author of his last scientific paper. "I look forward eagerly to when that fifteen pence will be gone."[6] He stopped eating anything except a large glass of milk every day and had to be hospitalized for malnutrition.

Alcoholics he had taken into his house stole what little he still possessed. Then Price plunged into periods of depression. Although he continued his research in genetics, he soon had a quarrel with his colleagues; an alcoholic whose wife Price had assisted vandalized the genetics institute.

In the fall of 1973, Price had to vacate his apartment. He gave away the last of his possessions and lived as a vagrant. He worked nights for a cleaning company and during the day volunteered in old-age homes. The director of one in which he showed up on Christmas Eve later recalled that it was "like an angel coming in."[7] Later, he found refuge in a squatter commune.

Toward the end of 1974, he decided that "Jesus wants me to do less about helping others and give more attention to sorting out my own problems," as he wrote to his mentor Hamilton.[8] The latter persuaded him to rededicate himself to his scientific work, but he had another crisis shortly before Christmas. Friends later reported that he was in despair at not being able to really help the homeless. Price was found dead in the building occupied by the commune on January 6, 1975. He had slit his carotid artery with a pair of fingernail scissors.

Evolution in All Channels

Although William D. Hamilton continued to champion Price's ideas, the latter's accomplishments in genetic theory were largely forgotten.[9] Hamilton's earlier conviction that altruism was advantageous only among blood relatives had become too firmly established.

Most biologists seemed not to notice that Hamilton himself had long since revised this opinion. As if hypnotized, in the following years they continued to search for evidence that evolution was a continual struggle of gene versus gene. This hypothesis explains altruism between relatives in the simplest possible terms: People sacrifice themselves for their family members for the sake of their own genes, not because they want to do their relatives good. The new science of sociobiology, based on Hamilton's early work, declared that altruism was really motivated by nothing but the egocentrism of the genes.[10]

Of course, sociobiology has given us some valuable insights. For instance, it has been able to explain how the close genetic relationship of worker ants promotes their self-sacrifice. This speculative discipline, however, has also encumbered the world with untenable speculations like those of the American biologist Randy Thornhill, who postulated the idea of a rape gene in male *Homo sapiens*. Thornhill claimed that this predisposition has survived in the course of evolution because its carriers engender more progeny. As proof, Thornhill adduced the rough mating habits of the scorpion fly.[11]

The assumption that evolution merely chooses among single genes is already questionable. In and of itself, a gene is powerless, nothing more than a piece of the genetic substance DNA, dead material that bacteria would devour in a flash. Genes need living organisms to survive in. And in the end, it is the protozoa, plants, animals, and humans who are the players on the stage of life. They compete for resources, reproduce, or become extinct. Only if a living organism is successful do its genes have a future. It's never a single gene that is inherited; what is passed on to posterity is a package of thousands and thousands of genes: the individual's entire genetic makeup.

Progress in molecular biology makes clear how completely divorced from reality is the idea that evolution is merely competition among individual genes. For the operations by which genetic information is read and processed are much more complex than researchers had expected. Just as a jumbled pile of toothed wheels is very far from constituting a clock, the genotype is more than the sum of its genes. The combined effect of genes working together and with other components of the cells is the decisive factor. Which characteristics disappear and which survive in the evolutionary process depends on cooperation among hundreds of thousands of individual building blocks.

The Price equation raises this idea to a higher level. It is not just that single genes are joined together in the genotype of an individual who must compete with other individuals. In an analogous way, many individuals together form a group and then together enter the competition for the best chances to reproduce. And that's how evolution works everywhere: Individual genes compete for their place in the genotype, but they also work together so that the individual organism has the best reproductive chances. Individuals compete for resources within the group, but they unite and work together against other groups.

Reshuffling the Cards

More and more geneticists and behavioral scientists are endorsing this view of evolution,[12] but not all are convinced—not by a long shot. One of the biggest areas of contention in biology is whether nature chooses only between genes or between entire organisms and groups. And the question is as bitterly contested as if it were a religion being defended. Some praise the principle of group selection championed by George Price as a long-overdue victory over individualism. Others pillory such ideas as unscientific nonsense and declare the group to be a myth.

But this quarrel is unnecessary. Despite all the polemics, there's actually no contradiction between the two positions. As the mathematically talented geneticist Hamilton had already proved in 1975, a few formulas can bridge the gap between the various locations where evolution occurs. The competition between individuals and groups can always be interpreted as the competition between genes.[13] Of course, one has to be precise about what "genetic relationship" means. Genetic relatives do not always have to be related by blood; they can also have different ancestors.

It was this apparent contradiction that Hamilton had overlooked in his early publications. For example, all people with green eyes are related to the extent that they possess the same genetic variation for the color of their irises. What counts as genetic relatedness is only the genetic makeup one carries, not where it comes from. And whether a certain genetic variation—the one for altruism, for example—survives depends on genetic relatedness alone. As Hamilton showed, selflessness has already triumphed in evolutionary competition when people with the appropriate gene stick together, even if they come from different families. That's exactly what George Price's group theory says as well.

Further calculations, however, led to the conclusion that group selection can only promote altruistic actions in the long run if there is a certain amount of genetic mixing. Within groups, evolutionary pressure works in favor of the egocentrics. Even if the groups with the most altruists have more reproductive success than other groups, the proportion of selfish individuals within the total population inevitably rises. Only if a certain amount of exchange is allowed—e.g., if a small band of people separates from a group that has grown too large and joins a different group or sets off on its own—will altruism spread throughout the entire population.

It is simplest to illustrate this somewhat complicated idea with the

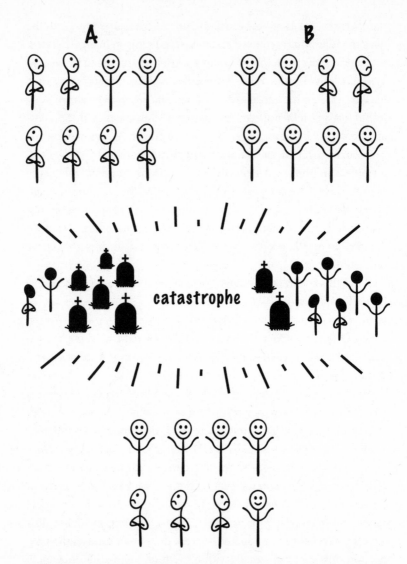

The good guys win out. To begin with, a population consists of equal numbers of selfish and generous people. In village A live mostly egocentrics, in village B mostly altruists (top of the page). When a natural disaster strikes, generosity wins out. Because six inhabitants of A, but only two inhabitants of B, die in the catastrophe (middle of the page), after it is over the altruists are in the majority. If the survivors are distributed between the two villages and reproduce, the game can begin again at some point. In time, the proportion of altruists will grow larger and larger through this "group selection."

example of two neighboring villages of equal size. The population of the first village is fairly selfish; only every fourth person will sometimes give food to others if his family has more than they need. In the second village, by contrast, it's exactly the opposite. There, three out of four inhabitants are generous. If you take both villages together, the population consists of 50 percent egocentrics and 50 percent altruists.

But now the country is hit by a severe drought. Since they stick together, the men and women of the altruistic village are better able to withstand it: Three-fourths of them survive. The egocentric village, on the other hand, is ill prepared for the ensuing famine: Only a fourth of them survive. Their proportion in the total population also shifts thereby. Now egocentrics make up only 37.5 percent, while 62.5 percent of the population are now altruists.[14] And since the less-selfish village will continue to make out better in future crises, the proportion of selfless people in the population will gradually increase.

The Right Amount of Generosity

While the altruistic groups weather the lean years better, the freeloaders come into their own during the good years. In the long run, no society can consist exclusively of selfless people. Sooner or later, the spongers and power-grabbers show up. It's a common observation that organizations founded by idealists later fall into the hands of petty functionaries.

In the worst-case scenario, the altruists die out; in the best case, a balance is reached between altruistic and less-generous behavior. The Price equation accounts for this dynamic as well. The tragic irony is that Price himself refused to accept the consequences of his insight and in the end fell into despair. His formula lucidly sorts out the two decisive forces: Competition between individuals favors egocentrism, while competition between groups favors selflessness. Thus altruism can only persist if altruists parcel out their service to others in the right dose. If they withhold too much, their sacrifice will not be effective and the community will be wiped out by other groups. But if they do too much good, they will fall behind vis-à-vis the takers in their own group. Over time, ruthlessness will win out, and in the long run, that community will also be destroyed.

With the help of the Price equation, it is even possible to calculate the right amount of generosity. For that purpose, the American economist and anthropologist Samuel Bowles used a model that is constructed like the competition for survival of the two villages described above, but closer to reality.[15] He collected data on the structure of prehistoric populations derived from archeological discoveries in excavated graves as well as from the way of life of contemporary small-scale societies and came to a surprising conclusion: Altruism will survive even if all the members of a society forego on average only 3 percent of their reproductive potential in favor of the community. Nor does it matter how this sacrifice of biological fitness is distributed within the group. If there were ideal fairness, all would contribute their 3 percent; but only every other person could be an altruist and donate 6 percent while the other half gave up nothing. In that case, the altruists in every generation would have 6 percent fewer descendants than their fellow members who didn't worry about the welfare of others. If there were no natural disasters, the altruists would die out in a few dozen generations. But since famines, floods, and wars with other peoples wreak less havoc on societies who stick together and help each other, altruism persists.

That this unending struggle between more selfish and more altruistic characters has left behind traces in the human genotype is suggested by the data that geneticists have collected at a unique festival. The small town of Twinsburg, Ohio, celebrates the first weekend in August every year by inviting twins (and triplets and other siblings from multiple births) from all over the world to a celebration that includes a carnival, a twins' parade, and a barbecue. Invitees include scientists hoping to enlist the twins to help them discover the genetic basis of human behavior. Special tents are erected where twins are asked to give blood and sputum samples, have their teeth measured, and fill out questionnaires about their lives.

The American political scientist James H. Fowler asked individual twins to play the trust game with another person. He was interested in whether twins were similar in their willingness to be trusting and to reward a favor. His results depended on whether the twins were identical or only fraternal. The identical twins were more like each other in their behavior. This was also the result reached by Swedish investigators who had twins play Ultimatum to measure their generosity and their willingness to punish others for unfair behavior even if it cost them something to do it.[16] Now that we have the data from 600 pairs of American and Swedish twins, there is no question that our genotype influences how much we are ready to do for our fellow humans. Altruism is also inscribed in our genes.

The Origin of Shame

It could be objected that having to give up only 3 percent of one's reproductive potential isn't very much. At first glance, it looks like the proverbial drop in the bucket. It's hard to believe that it could help a group achieve a significant competitive advantage.

But that overlooks the fact that altruists have a strong kind of leverage at their disposal: They can inflict punishment. As we've seen in the previous chapter, that's what changes the whole game. Sanctions make it expensive to help oneself at the cost of everyone else. In extreme cases, miscreants are banished from the community and left to fend for themselves in a hostile environment. So it's a better strategy to do one's part and not attract too much negative attention. And once the rule is established that every member of the group makes a contribution, the burden on the individual is not so great, since the weight is equally distributed on many shoulders. At the same time, the fitness of the entire community rises.

By inflicting punishment, the altruists among our forebears created a new evolutionary pressure. Individuals who hewed to the norm reproduced more successfully than the troublemakers. Thus life in the group had an effect on human genetics, and society began to put its stamp on our ancestors. Shame is an impressive instance of this interaction between genes and social behavior. There is no doubt that shame and contrition are inherent human characteristics, common to all cultures. They are evident in children before they become two years old.

But while almost all other human emotions are also to be found in animals, the present scientific wisdom is that shame and contrition are exclusively human, strong evidence for the assumption that they arose only by assimilation to life in a community.

Many dog owners believe that their pet is contrite when it has done something wrong, since it puts its tail between its legs and whines guiltily. But experiments prove that this apparent contrition is an illusion.[17] Actually, your dog is just cringing before the anger of its owner. If master and mistress are out of sight or don't seem to mind that their slippers are being chewed to pieces, dogs show no sign of contrition. Conversely, even when they haven't done anything wrong, they whine piteously when their owner scolds them.

However, observations of dogs do give us a valuable insight into the origin of shame. Clearly, shame grew out of emotions that regulate dominance and subservience, a conclusion further supported by the fact that even today, members of many Indonesian ethnic groups show all the signs of shame as soon as someone appears who is superior to them in rank, even if they have nothing to be ashamed about.[18] Just as a dog or a chimpanzee defers to the dominant animal, modern humans submit to the impersonal norms of the group.

The cynical belief is wrong: People follow rules and obey laws not just because they think themselves observed and fear punishment. How would you feel if your car had accidentally struck an old lady on a lonely country road? You could flee the scene. The victim would never remember your license number. But if you did, you could never look yourself in the face again. Shame is a punishment we inflict on ourselves, and thus an extremely effective early-warning system. Since we frequently cannot be certain that we're really unobserved, shame prevents us from kicking over the traces. We are programmed to live in a world full of rules.

"That's not how you dax!"

My two-year-old daughter insisted on brushing her teeth *before* she got undressed. Any attempt to alter the sequence was met by screams.

This behavior could hardly be attributed to compulsive subservience, since she usually responded to every suggestion from her parents with "Don't want to!" Her bedtime ritual became even more bizarre after I twice happened to be combing my hair while she was brushing her teeth. For weeks thereafter, she demanded that the two actions had to be done together.

Even hand puppets have to obey the rules. Hannes Rakoczy, a developmental psychologist in Göttingen, taught three-year-olds an invented game he called "daxing."[19] The point was to move a block using a rod and a pusher. After a child was allowed to dax alone for a while, a hand puppet appeared and said it would like to have a try. But as soon as the puppet dared to move the block without using the rod and pusher, the child would protest shrilly, "That's not how you dax!"

A norm is a rule enforced by the community. If you follow it, you're rewarded; if you violate it, you're punished. Seen in this light, the three-year-old subjects were trying to establish a norm at the play table: You dax with a rod and pusher. It was an impressive demonstration of how easy it is to establish a rule for human behavior. Developmental psychologists had always maintained that children either teach each other norms or internalize them out of fear.[20] But daxing is a one-person game, and no one had told the children that it was forbidden to dax without the pusher. And of course, there was no threat of punishment. But the simple demonstration of the game by the investigator was enough to implant a norm in the children's heads. The boys and girls did not have to play by the rules; they obviously felt an inner need to do so.

Like our ability to learn language, our willingness to adopt norms seems to be a basic part of our intellectual configuration. In fact, the two processes are similar. The development of the genetically determined brain structures begins in infancy, and the child absorbs what it experiences like a sponge. No coercion is necessary. It learns words and sentences and also behavioral rules all by itself. Its environment of course determines which language and which norms it acquires, but *that* it happens is innate. After all, both skills bestow incalculable advantages on social beings: They lessen losses caused by friction within the community and make it possible to pursue common goals.

The innate ability to learn rules makes it easy to teach people to behave altruistically. Along with our empathy, our innate willingness to help, and our ability to bond with others and adopt their points of view, this ability to learn rules belongs to the rich genetic endowment that enables human altruism. In the traditional societies of our ancestors, these predispositions continued to be reinforced. Helping others was rewarding; behavior that hurt the community was punished. In this way, humans cultivated their amicable characteristics. But nature provided the decisive impetus.

Not Genes Alone

To be sure, not every norm improves our life together. For example, after proposing a toast and lifting his glass to each person at the table while looking them in the eye, a real dyed-in-the-wool Bavarian must set down his beer glass briefly first before lifting it to his lips, taking a long swig, and emitting a heartfelt "Aaah!" It's not easy to make any sense of this ritual.

Other rules actually make interpersonal relations more difficult. Drivers in German cities who so much as touch the bumper of another car while parallel parking receive dirty looks and sometimes even a ticket. In other countries, by contrast, word has gotten around that that's what bumpers are for and that exaggerated concern for them causes unnecessary strife. To a Frenchman, the German parking rules are as foreign as the drinking customs of Bavarians are to someone from Hamburg.

But some norms are found in very similar form all over the world. The anthropologists who had people from widely varied cultures play Ultimatum seem to have stumbled on one of these universal rules. Although the members of some tribal societies had different ideas of what constituted a "fair division," first players the world over were willing to turn over more money to their unknown partner than economic rationality would have predicted. And the second players in all cultures were equally willing to reject offers they considered too stingy, even when it was to their own disadvantage.

Such behavior is very unlikely to be innate. Not until they begin school do children develop a marked sense of fairness; obviously, they have to learn to share by adopting the norm of what their environment considers fair.

Accordingly, peoples all over the world developed similar norms independent of one another. The rules of human communality thus seem to be anything but random. If we understand evolution in Hamilton's sense as a process on several levels, there is a plausible explanation for how norms developed. Just as people within a community compete for the chance to reproduce, entire societies compete for resources. The individual owes his biological fitness to his genes and his environment. Which *communities* survive, on the other hand, depends largely on the rules under which their members live together.

The History of Morality

Elementary questions such as "What is just?" would thus find their answers in a process that has taken millennia. This idea may seem unusual and even provocative. For usually we hear that our moral convictions are the result of conscious agreement, genetic determination, or divine revelation. Yet there is no empirical evidence for any of these origins. Thus the suspicion that our concept of good and evil crystalized in the course of a gradual cultural evolution is an attractive speculation at the very least.

It is supported by the observation that a people's shared conception of justice can be systematically traced to the circumstances of its communal life, as we have seen in the case of the generous Lamalera and the stingy Machiguenga. And the Erfurt experiment in behavior, also described in the preceding chapter, makes clear how these mechanisms work. Altruism became prevalent because participants voluntarily joined groups in which free riders could be punished, while the groups with lax norms were gradually depopulated. Thus it automatically became the universally acknowledged rule that one had to contribute to the community.

There are many examples of these processes to be found in history. The conflict between Athens and Sparta is a particularly dramatic example.[21] A comparison of the two republics shows how widely norms can diverge, even between communities belonging to the same civilization. Nor is it likely that there was much if any genetic difference between the populations of the two Greek cities. But while the Athenians cherished their personal freedom, in the militarized society of Sparta, the community was everything, the individual nothing. And these differences were decisive for the fate of the Greek city-states. For as the British historian Antony Black expressed it, the polis "was evolutionarily successful because members were on the whole willing to die for it."[22]

In Sparta, a relatively small circle of free citizens controlled the state; strict rules bound them together in a closed elite. For example, to even once show oneself cowardly in battle meant losing one's rights as a citizen. Such a blot was impossible to expunge even with subsequent heroic deeds. Getting into financial difficulties and not being able to pay one's share of the daily communal meal also meant demotion to second-class status. Most inhabitants were without rights in any case and little better than slaves.

In the Peloponnesian War, the extreme altruistic norm of the Spartans proved successful. This custom of selflessness contributed to the fact that Sparta and its allies were able to wage war for twenty-five years against the vastly superior economic power of Athens. In the end, the fleet of landlocked Sparta, built with Persian money, was able to defeat the fleet of the Attic naval power. Athens was forced to surrender in 404 BCE; Sparta assumed hegemony over Greece for several decades. But the disadvantages of the Spartan norm soon became apparent. The successful model of a small elite whose members were prepared to sacrifice themselves completely and were therefore highly effective in battle could not be expanded to fit an entire country. At the same time, egocentrics undermined the once-heroic Sparta, for the rich spoils of war created inequality. While some families amassed fortunes, others were impoverished and lost their social status. Although the exclusive circle of Spartan citizens was bleeding to death, the rigid social hierarchy remained in place instead of being relaxed to allow the intermixing necessary for the long-term success of altruism. Sparta's fighting capacity melted away, as did the powerful cohesiveness of its citizenry. When Philip II of Macedon, the father of Alexander the Great, conquered Greece, the advance of his troops put an end to Sparta's altruistic norm.

It has more often been the case, however, that norms evolve without violence. For one thing, societies learn from one another: When the rules of living together in a group and sharing prove successful, they are quickly adopted elsewhere. For another, prosperous communities have always had a powerful attraction everywhere in the world. Whoever wanted to join them had to live by their rules. Thus more and more humans adhered to the successful norms, while the less useful norms were gradually forgotten.

The Price equation can also explain such processes. As Price himself soon realized, his equation is equally valid whether altruism is spread genetically or via social norms. It depends only on the costs and benefits of selfless behavior for the individual on the one hand and for his group on the other, as well as on the composition of the groups. And in either case, strong competition between groups promotes altruistic behavior.

It's even possible to chart the success of entire religions and the moral development of their adherents using the Price equation. There is no doubt that the early Christians (as well as the Muslims in India, for another example) gained adherents because there was more solidarity within those groups than in the surrounding society in general. Whoever joined them could depend on his brothers and sisters in faith. They all strove to attain the ideal stated in the epigraph from Acts at the head of this chapter. But how were these communities of believers to continue to distinguish themselves when their faith became the state religion? Although—or precisely because—all of Western Christendom now professed to follow the norms of the Sermon on the Mount, customs became prevalent within the church that were worse than those at the court of the Roman Caesar. The group had lost its competition.

CHAPTER 9

The Evil in Goodness

And the whole earth was of one language, and of one speech. . . .
And they said, Go to, let us build us a city and a tower, whose
top may reach unto heaven . . . And the LORD said, Behold,
the people is one, and they have all one language; and this they
begin to do: and now nothing will be restrained from them,
which they have imagined to do. Go to, let us go down, and there
confound their language, that they may not understand one
another's speech. So the LORD scattered them abroad from thence
upon the face of all the earth: and they left off to build the city.

GENESIS 11:1–8

MUZAFER SHERIF WAS TWELVE YEARS OLD WHEN soldiers landed on the beaches of his home town of Izmir in Turkey. The armed men spoke Greek and called the town Smyrna, as did half the citizens of the town itself. The Greeks in the city rejoiced when the occupiers entered the town, but soon there were scenes of horror. Before the bayonets of the soldiers, the Muslims of Izmir were forced to take off their traditional fezzes and trample them on the ground while cursing against Moham-med. Whoever refused was murdered. There were massacres on that fifteenth of May, 1919, as reported later by a commission of inquiry appointed by the Americans, English, French, and Italians. The Greek soldiers tied the hands of Turkish-speaking men, women, and children, lined them up, and shot them.[1]

Muzafer Sherif found himself in one of these doomed lines. One victim after another was shot. When the man next to him fell, Sherif knew his turn was next. The soldier was loading his rifle. But suddenly he hesitated. Did he feel pity for his young victim? Sherif never learned why he was spared, but the Greek soldier turned away and left.

In the following months, in Izmir as everywhere in the disintegrating

Ottoman Empire, the separate ethnic groups fought one another. Greeks persecuted Turks, Turks persecuted Armenians, Christians burned down Muslim houses, and Muslims murdered Christians. It was not so much the violence itself that puzzled Sherif, however. What made a deep impression on him was the selfless comradeship in each group. The altruism within the community corresponded to its hostility, murderous rage, and vengefulness toward its hated enemies. People were simultaneously sympathetic and possessed by prejudice, ready for both supreme self-sacrifice and bestial acts. What was going on in their heads? Sherif decided to devote his life to investigating and understanding the causes of this paradox.[2]

The Eagles Versus the Rattlers

Sherif's experiences in Germany twenty-two years later must have seemed like a nightmare returned from his childhood. The young social scientist had gone to Berlin to study psychology, but goon squads from the SA, the SS, and the Communists were battling one another and the police in the streets. Hitler was named chancellor in 1933 and began persecuting his political opponents and Jews in the name of the *Volksgemeinschaft*—the community of the German *Volk* (people).

Sherif emigrated to the United States, where in June 1954 he conducted a groundbreaking experiment.[3] He and his wife, who collaborated on the experiment, had spent weeks trolling schoolyards for appropriate subjects. They were looking for eleven-year-old boys who were, from their perspective, completely "normal" and average in every respect—"healthy, socially well-adjusted, somewhat above average in intelligence and from stable, white, Protestant, middle-class homes"— but unknown to one another.[4] Once they had found twenty-two such boys, they divided them into two groups of eleven and had them picked up by two separate busses.

They were driven to a summer camp on a river. At first, the two groups were kept far apart, without any contact. Sherif posed as the director of the camp and documented everything that happened. After less than a week, each group of eleven boys had become a strong community. They had even given their "tribes" names; one group called itself the Rattlers, the other the Eagles. In each group, leaders and followers had emerged and the boys had developed rituals and customs. The Rattlers specialized in profanity. The Eagles went skinny-dipping.

Sherif had planned to induce conflict between the groups, but that proved to be unnecessary. Although neither band had previously laid eyes on the other, some boys were already calling the other group "nigger campers" at the mere sound of their voices in the distance. They asked their adult counselors to organize a competition with the strangers as soon as possible. At the same time, the audible presence of the other group altered the behavior within each group. The boys helped each other and shared more. Non-swimmers who had been looked down upon were now actively helped to learn how, "so that we'll *all* be able to swim."[5]

When Sherif brought the two bands together at last for group competitions such as rope-pulls and treasure hunts, they reviled each other as "communists," "stinkers," and "sissies."[6] Although none of them had the least reason to be hostile, the Eagles and the Rattlers apparently couldn't stand each other. They held their noses when a member of the other group was nearby and refused to eat together.

On the next day, the Eagles burned a Rattler flag; the Rattlers retaliated by attacking and destroying a hut the Eagles had built. They found a pair of jeans belonging to one of the Eagle leaders and raised it on their flagpole with the motto "The Last of the Eagles." Thereupon, the Eagles armed themselves with baseball bats and ransacked the Rattlers' camp. Luckily, the other band was not at home.

The more bitter the rivalry became, the stronger the intra-group selflessness grew. Especially, the youngest boys at the bottom of the pecking order tried to distinguish themselves through heroic action—and with special hate for their enemies.

Attempts by Sherif and his colleagues to overcome the rivalry were unsuccessful. Neither a joint trip to the movies nor fireworks on the Fourth of July was able to bring Rattlers and Eagles closer together. A meal planned to seal their reconciliation ended with a food fight in which the boys threw steaks and chicken drumsticks at one another.

We will see below that Sherif did succeed in finding a way to make peace, but the conclusion he drew from the study was still unsettling: Groups hate and attack each other not because psychopaths like Hitler seduce them into it. Conflicts between communities that seem senseless to an outside observer arise spontaneously. All one has to do is bring two average citizens together; without further prompting they will begin to reject outsiders and find reasons why they are different from others.

In direct proportion to the increase in hostility toward outsiders, selflessness and solidarity grow within a community. Israeli investigators organized a competition between harvesting teams in the orange groves.[7] Although each group continued to be paid as a whole and members therefore had a strong motivation to let others work harder, no one dared to try freeloading anymore. The average yield per group rose by 30 percent.

Corruption through Selflessness

Does this mean that xenophobia, vindictiveness, and war are the price we must pay for our ability to act altruistically? After all, leaders have always instigated conflict to unite their subjects behind them.

It lies in the nature of selflessness that cooperation needs competition, since altruists are under constant threat of being exploited. If in the long term they give too much to the wrong people, it will lead to their downfall. They can only survive by distributing their charity unequally. They should confer their largesse preferentially upon people like themselves—other altruists. As long as the selfless stick together, they all benefit from the cooperation and are thereby at an advantage over egocentric loners. But if they help others indiscriminately, their charity goes for naught: Random people profit from it and are happy, but they give no thought to doing something for others themselves.

The difficulty, of course, is finding the right recipients. It's simplest to stick to the people you know and trust—your friends. Entire careers have been built on the principle "one hand washes the other." Almost one-third—and in small businesses and particular industries like financial services even more than half—of all jobs in Germany are filled through personal contacts.[8] It's practical for all concerned and sounds harmless enough. But of course, it is to the detriment of applicants who don't know the right people as well as of firms that aren't hiring the best-qualified candidates.

Altruism and morality are two different things. People who give their feckless buddy a leg up are often behaving with true selflessness. If a man helps the not-very-bright son of his ex-girlfriend get a job in his company, he's not doing it with the hope of getting anything in return, but out of old affection. Yet he thereby damages the company.

Of course, many generous people hope that their actions will be repaid some day when they themselves are in need of a small favor. But usually, repayment comes not from the particular recipient of their generosity, but from someone else. That's the principle of networking. Whoever belongs to such a group can count on the support of the others, but also has to be there for them. A group of politicians in the conservative Christian Democratic Union in Germany formed such a network during a trip to South America in 1979. The members of the group, which became known as the Andes Pact, pledged never to say a word against another member, much less oppose one another in party primaries. Many of them rose to prominence and held dominant positions in German politics for decades.

Their success is a perfect illustration of the principle of group selection described in the previous chapter. A clique also depends on selflessness. If an ally needs support, it must be given. Whether the supporting person benefits thereby doesn't matter. It is the group norm; whoever refuses is ejected from the group. Several members of the now-defunct Andes Pact made the unpleasant discovery that membership in such a network does not automatically benefit everyone in it. While some rose to powerful positions with the group's support, others languished in provincial obscurity.

A study of small-scale societies in Papua New Guinea shows how weakly altruism is connected to higher moral principles. Swiss investigators had members of two mutually hostile ethnic groups take part in experiments that involved sharing.[9] The subjects only cared about fairness with respect to their own community. If a member of the enemy group was swindled, they were less ready to punish the miscreant, and they were unwilling to punish him at all if he came from their community. Those Papua New Guinea hunter-gatherers are not so different from some trade unions in democratic societies: They fight for their members but don't do much for the unemployed.

Even the success of the Mafia depends on extremely altruistic group norms. Whoever joins must submit to its infamous code of *omertà* and keep silent even if he is unjustly accused of a murder. According to a list of "Ten Commandments" that the Sicilian police discovered in November 2007 in the hideout of the captured boss Salvatore Lo Piccolo, the mafiosi must put the interests of the organization above their own in every respect, "even if their wife is in labor." A good mafioso must never lie or take money "that belongs to others or their families." All this assumes, of course, that the "others" are also "men of honor."

"When Envy Breeds Unkind Division"

It is characteristic that the codex of the Mafia, like the rules of the Andes Pact, places great emphasis on group members not competing with one another. For the more a group is able to direct the competition for resources outward, the more successful it is and the more intramural altruism flourishes. Instead of competing with one another, group members attempt to capture a bigger piece of the pie together at the cost of other groups.

This situation arises spontaneously whenever competition within the community is unproductive because everyone is equal. And that seems to be the state in which our ancient forebears did indeed live.[10] Almost all hunter-gatherer societies that have survived into modern times and have been studied are egalitarian. Whether on the steppes of East Africa, in the jungles of the Amazon, or in the highlands of Papua

New Guinea, important decisions are as a rule made communally and members of the community react strongly when one individual tries to rise above the others. In contrast to our nearest biological relatives, the great apes, hierarchies are absent among the hunter-gatherers. If someone tries to make himself into a boss or accumulate too many possessions, he will face stiff resistance and possible exclusion from the community. There was obviously strong evolutionary pressure on our forebears to resist all-too-noticeable differences in status for the sake of cooperation in the group.

Thus the longing for equality seems to have become a part of our nature. It plagues us to this day in the form of envy, a feeling that exists in no other animal—the wish to take something away from someone else to bring that person down to the same level. Envy is painful, as measurements of brain activity show. To see someone else in a better position than oneself activates similar processes in the brain as when one is stabbed with a knife.[11]

Thus it's not surprising that envy has such explosive potential. Studies in social psychology have repeatedly shown that people are willing to forego part of their salary as long as others get even less. But the gnawing feeling of envy all too often breaks out in psychological harassment, damage to property, and even physical violence.[12]

"When envy breeds unkind division / there comes the ruin, there begins confusion."[13] In *Henry VI*, Shakespeare sums up the double-edged nature of envy. In its destructiveness, it often injures the envious person as much as the person envied. Selfless actions are advantageous from an evolutionary standpoint as long as the community wins at the expense of its altruists. But when farmers in some remote corners of Eastern Europe even today set fire to the farm of a neighbor who they think has become too prosperous, on balance they lessen the future prospects of their community.

Such acts can only be explained as relics from archaic times. In a tribal society whose disintegration would mean the ruin of all its members, any hint of inequality is an existential threat. To forestall that threat, it can be worth the cost to temporarily reduce the resources of the entire group.

Deadly Taboos

It is not just envy that causes communities to make dangerous sacrifices in order to stay together. Dietary taboos restrict the amount of available food. The higher castes in India, a country often plagued by famine, may not eat meat and thus leave an important source of protein unexploited. Even more remarkable is the discovery that the Greenland Vikings left no traces of having eaten fish. Archeologists working in their abandoned habitation sites have found no fish hooks, net weights, or other traces of fishing. Could it be that this seafaring people on the coasts of Greenland really ate only seals and the few livestock able to survive north of the Arctic Circle? The geographer and evolutionary biologist Jared Diamond suspects that this was in fact so; a taboo forbade the Norsemen in Greenland from eating fish, in contrast to their relatives in Iceland and Scandinavia. And so they were unable to survive in their arctic environment when the climate became considerably colder. The less-choosy Inuit, on the other hand, survived the harshest times.[14]

Many of the initiation rituals customary in almost all societies seem to us destructive and cruel. In some African and Oceanic societies, adolescents are inflicted with deep cuts into which ashes are then sprinkled so that they take as long as possible to heal and leave behind so-called decorative scars. Apart from the pain, there is also a high risk of infection. Many peoples perform such rituals, especially on the genitalia. In some cultures in Oceania and among Australian Aborigines, boys entering puberty have their penises cut open. In Indonesia, bamboo balls are implanted in their penises. In parts of Africa, female circumcision is still widely practiced. It not only entails excruciating pain but also deprives many of its victims of sexual pleasure for the rest of their lives and raises the risk of death at delivery for both mother and child. Under the usual extremely unhygienic conditions, the operation itself is often life-threatening. A people with such traditional customs not only does violence to individual women and men but lessens its own biological fitness as a whole.

Why do people inflict such things on themselves? Dietary taboos, voluntary mutilation, life-threatening initiation rituals, and complicated sacrifices all serve not only to strengthen the community's cohesiveness but to differentiate it from other groups. The form of the decorative scarring almost always indicates what group the bearer belongs to; once scarred, that person remains identifiable as a member of a particular clan or people. Even dietary taboos are effective in keeping people divided, since we acquire our tastes for food early in childhood and disgust is such a strong emotion. It's not just that Europeans and Americans would never touch the meat of a dog specially bred to be butchered and eaten; the very thought of Korean gourmands enjoying this delicacy makes us nauseated.

Rituals and taboos reinforce the boundary between the members of a community and "the others." They insure that the fruits of selfless action benefit only one's own people.

A study by the American anthropologist Richard Sosis shows the degree to which costly proscriptions hold a group together. He analyzed the fate of 200 communes that experimented with new lifestyles in nineteenth-century America.[15] Some of the proto-dropouts were the adherents of socialistic or anarchistic ideas, while others lived according to their religious convictions. Sosis compared how long the communities survived and discovered that the religious communes had four times as much chance of surviving as the secular ones. Moreover, the stricter the former were, the more successful they were. Precisely because fasting, celibacy, and abjuring private property and modern technology lessen each individual's chance of surviving, they promote the success of the community.[16] Of course, the sacrifice of biological fitness must not exceed a certain limit—must not be *in toto* higher than the accrued benefit to the community. Otherwise, as the example of the Greenland Vikings shows, the self-imposed limitations can easily lead to disaster.

The Tower of Babel

There are more than 800 languages still spoken today on Papua New Guinea, some as different from each other as English and Chinese. If one believes the story in Genesis with which we introduced this chapter, the babel of languages arose as God's punishment for humanity's hubris. Never again should there be an effrontery like the Tower of Babel, and so the Lord "scattered them abroad . . . upon the face of all the earth."

. The authors of the biblical text rightly guessed that there was a connection between the lack of unity among humans and their greatest common achievements. But in the light of our present knowledge, we would tell the story in a different way: Cooperation on a large scale was first made possible by humanity's division into groups. For it is only under the pressure of competition between communities that individuals subordinate their private interests to the common good. Accordingly, the confusion of languages came first, then the Tower of Babel. Only the division of humans into separate groups enabled the building of the Egyptian and Mayan pyramids, the Acropolis, and finally the skyscrapers of Manhattan, Shanghai, and Dubai.

Fortunately, modern societies no longer express their identity through sacrifices or mutilations, but it is obviously part of human nature to unconsciously distinguish friend from foe on the basis of language. A 2007 study found that infants were already able to distinguish between people of their own nationality and foreigners.[17] By the age of five months, they preferred people who spoke their mother tongue. For example, when a baby raised in a German-speaking family heard a stranger speaking unaccented German, it would later seek more eye contact with that person than with another stranger whom it had heard speaking a foreign language. Although the five-month-olds did not yet know the meaning of words, they had already internalized the typical sounds and combinations of their native language.

And it is not just that the babies preferred familiar sounds. They reacted more positively to the native speaker even when that person had remained silent for quite some time. They recognized and rejected someone speaking German with an accent. At ten months of age, they preferred to accept a stuffed animal from a person who spoke without a foreign accent rather than from someone who had not learned German until she was an adult. And yet the foreigners spoke fluent German with only tiny variations in accent. The *r* of a Russian woman may have sounded too rolled or the *ch* of a Greek man too harsh, although a far greater proportion of their sounds were familiar. But the babies seemed to react to precisely these subtle differences—as if they possessed from the first a sensor for foreignness.

Normal Madness

Even in the trivial situations of everyday life, we can see how xenophobia is connected to our natural predisposition to be altruistic. For those of us who regard ourselves as cosmopolitan, it is depressing to observe that people will do more for a stranger who speaks their own language or at least comes from the same culture. Investigators have found this bias, for example, among Belgian university students.[18] The Flemings speak Flemish and come from northern and western Belgium; the Walloons speak a French dialect and live in the southern and eastern parts of the small country. When these highly educated young people played the trust game, the Flemings would offer less when they knew that their otherwise-anonymous partner was a Walloon, and vice versa. Of course, such behavior is neither a Belgian specialty nor a question of language alone. When the Israeli economist Chaim Fershtman tried the same experiment in Jerusalem, he had similar results. Strictly observant Jews were willing to share only with other orthodox Jews. If they knew nothing about the religious affiliation of the other player, they kept their money in their pockets.

At the extreme end of the spectrum are the crimes people commit in the name of their communities. Contrary to what one might expect, the attackers of September 11, 2001, were not characterized by a special inclination to violence or perversion. By now, entire archives are full of psychological analyses of the conspirators who flew the planes into the World Trade Center and the Pentagon. What those studies brought to light were unremarkable biographies. Ziad Jarrah is a typical example. He took over the controls of United Airlines Flight 93. Like the other three pilots, he had studied in Hamburg.[19] He came from a well-to-do family in Lebanon. His parents had sent him to a Catholic school. As a young man, he did volunteer work with handicapped children, drug addicts, and in an orphanage. He began his studies in Greifswald, Germany. He lived with his Turkish girlfriend and planned to get married. He transferred to Hamburg, where he had such a good relationship with his landlady that she painted an oil portrait of him. His life only changed when he came into contact with radical Islamists at the technical university in Hamburg-Harburg. He joined them and swore a blood oath to Osama bin Laden in Afghanistan. But during the whole time he was being trained as a terrorist, he kept the tie to his fiancée; on September 10, 2001, he wrote her a moving letter of farewell in which he thanked her for their five years together.

Analyses of Palestinian suicide bombers and of assassins who survived also reveal nothing remarkable.[20] The killers-to-be have normal personalities and are not even more religious than average. Another study in the Gaza Strip concluded that at most the militant young men were characterized by their pride, strong attachment to the community, and social consciousness. Almost all of them had provided supplies to other activists, visited the families of those killed, and taken care of the injured.

The feeling that one's own community is being treated unjustly makes young people susceptible to radical ideas, according to the American anthropologist Scott Adams in a study of suicide bombers. But they become involved in extremist organizations via friends or relatives.[21] And these bonds play a decisive role in their subsequent criminal careers. For organizations like Al-Qaeda and Hamas mold the close friends into conspirators and reinforce their hatred of the enemy. The often charismatic trainers consciously exploit the young people's predisposition to help others. They train together in three- to six-person cells until they take the final pledge to die for the cause.

A world populated by egocentrics would contain no one prepared to kill innocent people along with himself in the mad conviction that it would benefit his community. As early as the Book of Judges, however, we read of this most terrible manifestation of altruistic behavior. The Israelite hero Samson killed 3,000 enemy Philistines—men and women—along with himself when he knocked down the pillars supporting their banquet hall, "So the dead which he slew at his death were more than they which he slew in his life."[22]

Even more victims than in that ancient massacre or the attacks of September 11 were killed by the suicide missions of Japanese kamikaze pilots during the Second World War. Those insane acts were also driven by a strict altruistic norm. When the Japanese admiral Takijirō Ōnishi was looking for pilots for the first kamikaze missions in 1944, all twenty-three young airmen assembled to hear him speak raised their hands to volunteer. None wanted to be branded a coward. And it continued to be mostly young pilots who preferred to sacrifice themselves rather than have shame heaped on their heads. On April 6, 1945, Ōnishi's squadron carried out what was probably the most lethal suicide mission in history. About 1,500 planes appeared in the sky over the southern Japanese island of Okinawa and dived toward the American and English warships cruising off the island. More than thirty ships were sunk or suffered serious damage. It was this attack on Okinawa, among other things, that led American strategists to decide to use the atom bomb to break the Japanese resistance once and for all. A few days after the capitulation, Admiral Ōnishi asked forgiveness from the families whose young men he had sent to their deaths and then committed suicide himself.

A Common Goal

Even if the dark sisters of selflessness seem to follow a compelling logic, we are able to resist them. In the miniature world of his summer camp, Muzafer Sherif discovered an elegant trick to make peace between the Rattlers and the Eagles: He steered their aggressive impulses toward a new and common goal.

First, he secretly had the pipe that brought drinking water to the camp blocked up. When the boys learned of the emergency, the director told them that he needed more than twenty helpers to solve the problem quickly, since the entire line between the water tank and the camp needed to be checked. Rattlers and Eagles worked together to test the line, even lending each other tools. But as soon as the water was flowing again, they returned to their old habits.

Now Sherif organized a movie night for which both groups had to contribute toward the cost of renting the film. After that, he planned an overnight camping excursion for the whole group, but secretly disconnected the starter of the truck that was supposed to transport their food. And again, one team was not sufficient to get the vehicle started. As soon as the Eagles and the Rattlers realized they needed each other, they all pushed the truck to get it started. When they reached the campsite, it turned out that somebody had mixed up the components of the tents. To set up camp, members of both groups had to sort out tent pegs, ropes, and poles together. And when the food truck finally arrived, the main item was a nine-pound side of beef that had to be cut up.

All this led to conciliation. Rattlers and Eagles celebrated their last night of camp together, boarded a single bus for the ride home, and when the Eagles had no money left at a rest stop, the Rattlers treated them all to milkshakes.

Today, Sherif's idea is regarded as the classic path to reconciliation. If it is possible to give competing groups a common goal, they can overcome their rivalry. Former enemies then regard each other as allies in a new, common struggle. The eternally quarreling regions of Italy regularly forget their regional pride as Lombards, Tuscans, Sicilians, etc. and their contempt for all other regions whenever the national soccer team plays in the European Championship or the World Cup tournament. For two weeks, they're all just Italians and they all root for their team against the French, Brazilians, Germans, etc. (Once the cup is won or lost, on the other hand, the whole country reverts to its usual regional rivalries.)

The difficulty for a peacemaker in a real conflict is not only finding a common goal. Unlike Sherif, whose boys were not burdened by preexisting animosity toward one another, the real intermediary must overcome persistent psychic obstacles—memories of injustices that were suffered decades if not centuries ago.

But there is no lack of encouraging examples. Today it seems almost unreal that only a few decades ago, French and German troops lay in trenches shooting at each other, but in fact, the ancient enmity was overcome a few years after the Second World War. The rise of the Soviet Union as an atomic power contributed to the reconciliation. The fear that the Red Army could overrun the whole of Europe in a few days bonded Western Europeans together. But it was also trade, economic ties, and tourism that strengthened their common interests in the following decades. Later, the growing global competition for jobs, markets, and raw materials led to the realization that Europeans could only hold onto their high standard of living by working together. No one today can imagine that Germans and Frenchmen, Britons and Spaniards would ever again be capable of what had been one of their main preoccupations since the Middle Ages: trying to kill one another.

CHAPTER 10

The Golden Rule

Love your neighbor, for he is like you.
MARTIN BUBER

WITHOUT HER HELPERS, INGE DEUTSCHKRON WOULD NEVER have made it to her twenty-third birthday. But more than a dozen people stuck by her. They all knew the risk they were running by giving the young woman work, a bed, and some of their own meager rations—they shared her fears with her. And so Frau Deutschkron, at eighty-five still an extremely elegant woman, is able to sit in Berlin today and tell stories about the people she owes her life to.

She has returned to the place where she once went into hiding. It's one of those commercial rear courtyards typical of the German capital. No stranger to Berlin would guess they are there behind the splendid residential facades. They are so hemmed in by walls that no shaft of sunlight falls on them. Here hordes of workers once toiled in small workshops, one of which employed blind workers and was run by a man named Otto Weidt. The shop had a contract to make brooms and brushes for the Wehrmacht, the German army—quite a profitable business during the Second World War. Weidt was from a working-class family and had himself gone blind. When the Nazis began to require Berlin Jews to perform forced labor, he also employed Jews but treated them well. And so Inge Deutschkron applied for a job in the workshop. "I actually have no work for you," Weidt told her. The Nazis had prohibited Jews from doing office work. Nevertheless, Weidt hired the young woman and provided her with false papers.

Jewish Berliners were given less and less food as time went by, so Weidt also purchased groceries for his thirty-five employees. Since his shop was designated *kriegswichtig*—important for the war effort—his workers were protected from deportation for a while. But when the Gestapo finally took his blind and deaf workers off to a collection point in 1942, Weidt went to the Gestapo main office and apparently bribed the right people. A few hours later, an eyewitness saw a strange caravan wending its way from the urban rail station at Hackescher Markt back to the Rosenthaler Strasse. "The brush makers with the yellow stars on their chests, holding tight to each other, supporting each other, and in front was Otto Weidt, leading them back to the workshop."[1]

His workers called him "Papa Weidt," and he now began to arrange as many hiding places for his charges as he could find. He hid some in an earlier workshop in another part of the city. For others, he rented a shop he said was a satellite warehouse. Even in his main shop he created hiding places. Only a stone's throw from the Berlin Gestapo headquarters, Jews were living in his furnace room, and a four-person family found refuge in a windowless room he had partitioned off from his factory floor. In addition, a circle of supporters had been formed. They also took in Jews, forged papers, manipulated deportation lists, or visited the hideouts to provide medical treatment. Even some police officers from the station on the Hackescher Markt stamped forged IDs and warned Weidt of Gestapo sweeps of the neighborhood.

Inge Deutschkron and her mother began an endless series of moves. They lived in the back room of a laundry, the shed of an allotment garden, a worker's cramped apartment and a school principal's spacious one, a boathouse on the Wannsee, and a former goat shed. The owner of the goat shed was the only one who didn't know who the sleep sofa that Herr Weidt's truck delivered to her door was for. All the others who gave refuge to the Deutschkrons knew precisely the danger they were running along with their guests. Other helpers gave the two women work. Inge Deutschkron worked as a clerk in a stationer's while her mother did ironing in a laundry.

Through their courage, every one of these people helped save Inge Deutschkron and her mother from the death camps. Not all the Jews who Weidt strove so hard to save were as lucky. After things had gone well for eight months, an inhabitant of the windowless room ran into an old acquaintance on the street and naively told him everything. But the man was an informer. Two days later, the Gestapo showed up and deported all the inhabitants of the workshop. The police kept their hands off the head of the company himself; they probably feared he would reveal that they had taken bribes. Weidt immediately began to send food packages to the concentration camps. A total of 150 of his packages made it to Theresienstadt in the following months and actually reached the addressees.

What happened next sounds incredible but is fully documented. When Weidt's former secretary and lover Alice Licht was transferred to Auschwitz with her family, she threw a postcard out of the moving boxcar in which they were being transported. The card, still in existence today, is addressed to Otto Weidt and includes the note "Recipient will pay postage due." It was delivered. Under the pretext of wanting to sell brushes to the camp, the blind Weidt traveled to Auschwitz. When he learned that the young woman had been transferred to a camp near Wroclaw, he followed her there. He got a Polish worker to smuggle a message to Alice Licht that he had rented a room for her in a town close by and stocked it with clothes, medicine, and cash. He told her to try to escape the camp and use the provisions to make her way back to Berlin. And she succeeded in doing just that. Shortly before the approach of the Red Army, she managed to escape and get to Berlin. She hid out in Weidt's apartment. Inge Deutschkron and her mother were also still in Berlin at the end of the war.

While the survivors emigrated, Weidt stayed in Berlin and worked on rebuilding an orphanage for Jewish children, a refuge for boys and girls who had lost their parents in the camps. He died in 1947.

"They couldn't stand seeing people being sorted."

I asked Inge Deutschkron what she thought motived Weidt and other rescuers to their extraordinary actions. Her answer was short and to the point. "They couldn't stand seeing people being sorted." You have to think about that sentence awhile before you grasp what it means.

For Inge Deutschkron, what's unusual is not so much *that* people put themselves in danger to save others from certain death. What's remarkable for her is *for whom* Weidt and his friends risked their necks. Almost everyone would have tried to protect their own children or other close relatives from the Nazi threat. Most people would have also run the risk for very close friends. A surprisingly large number of people are even willing to risk death to save members of their community. That all seems to be part of human nature. But as described in the previous chapter, such altruistic sacrifice for the community is usually accompanied by rejection of outsiders.

The altruism of many of the people who helped Inge Deutschkron survive can in fact be understood as solidarity within a group. Deutschkron's father was a high school teacher and active in the Social Democratic Party. His party comrades supported his wife and daughter for years after he succeeded in escaping to England but the two women failed to obtain exit permits.

But by no means was everyone who came to their aid a political comrade. Nothing in Deutschkron's previous biography connected her to Otto Weidt or to Frau Gumz, the owner of the laundry. She was the first person to offer a hiding place to Deutschkron's mother, in November 1942. She was motivated by the stories of a neighbor who returned from the eastern front to say that the Nazis were gassing Jews in Poland.

The altruism of these people reached beyond the boundaries of their own community. It didn't seem to matter to Frau Gumz or Otto Weidt that the Deutschkrons had other political loyalties, that they were intellectuals from a different social stratum, or that they belonged to a different religion than their own. Nor did they allow themselves to be influenced by the measures taken by the Nazi regime to separate the

Jews from other Germans. That was precisely the point of the yellow Star of David that Jews were forced to display on their clothes; their exclusion from sports clubs, cafés, and streetcars; and—especially odious—the prohibition of selling soap to Jews. The Nazis wanted people to feel loathing for Jewish men and women by compelling them to be dirty.[2] Gumz, Weidt, and their allies, however, saw the Jews not as "the others," but as human beings who needed their help. They were unable to stand idly by.

Born to Be a Rescuer?

Is there such a thing as an altruistic personality? We owe the most thoroughgoing investigation of that question to the sociologist Samuel P. Oliner. His life's work has been to find out what personality traits and what life histories make some people into helpers while others look away. The organized mass murder of European Jews offered him material for his research with which he was intimately familiar. Although he lives in California today, Oliner is a Holocaust survivor from Poland. His parents were murdered in the Nazi death camps, but friends of theirs hid him when he was a child.

Oliner and his wife Pearl proceeded very methodically. In decades of effort, they sought out 230 people who had saved Jews from the gas chambers and conducted in-depth interviews with them about their own biographies. Then the couple compared the information they had gathered with the biographies of an equal number of contemporaries who had been in a position to help Jews but did not do so.[3]

The Oliners were looking for a pattern that would explain everything, but what they found were almost as many different motivations as there were interviewees. Some of the helpers had been shocked by the brutality of the persecutors or by the sight of a victim. Others were political opponents of the Nazis. For others still, the persecution and deportation of the Jews violated their religious or moral principles. And the biographies and personalities of the helpers were just as varied as their motivations.

The comparison of the helpers' answers with what the non-helpers told the Oliners was completely confusing. Why would one Christian run a great risk to help a persecuted person while another remained indifferent? Why did the same horrific scenes prompt some eye-witnesses to help but not others? The women and men who put themselves at risk for the persecuted were not particularly distinguished by special character traits or unusual life histories, nor were they especially empathetic or intelligent. The Oliners had to admit that they had been pursuing a will-o'-the-wisp. There is no rescuer personality. "Extraordinary Acts of Ordinary People" is the subtitle of one of his articles.[4]

There were only two points at which the Oliners could discern something unusual. First, people who would later become helpers very often identified with a parent who had strong moral norms. They obviously became accustomed early in life to act according to ethical principles and to be largely indifferent to the opinions of others outside the family. The second difference may be partly dependent on the first, but was even more marked: These future heroes found it especially easy to see themselves in a stranger's shoes. While the people who failed to act tended to emphasize that the Jews had lived differently than they did, the rescuers spoke more of their own similarity to the persecuted. They simply did not clearly perceive the supposed differences between Jews and Gentiles. In their opinion, if people needed help, they should get it. For them, right and wrong did not depend on what community the needy person belonged to.

The Love of Confucius

Measured against the 150,000 years of human history, the thought that moral principles should apply to everyone is very recent. Among the first to enunciate that position was Confucius. At the beginning of the fifth century BCE, he made a concept he called "ren" the core of his teaching. The word is usually translated as "humaneness," but Confucius himself, when asked what it meant, described it as "love" in the sense of "Love thy neighbor."[5] The character for "ren" is composed of two simple ideograms: The first means "person," and it is followed by two horizontal strokes symbolizing the number two.[6]

Confucius had a pragmatic reason why one should be led by the ideal of loving one's neighbor. When his pupil once asked him what "ren" meant for a ruler, he answered "When abroad behave as though you were receiving an important guest. When employing the services of the common people behave as though you were officiating at an important sacrifice. . . . In this way you will be free from ill will whether in a state or in a noble family."[7] Less than two centuries later, the philosopher Mengzi expanded Confucius' concept. He explicitly declared that every human being had the same basic rights. "He argued that humaneness, rightness (*yi*) and propriety (*li*) are rooted in the human mind. Knowledge of right and wrong and feelings of compassion and shame are 'possessed by all human beings.'"[8]

At about the same time as Confucius, in the fifth century BCE, a similar idea was being written down in the Near East. The authors of Leviticus left no shadow of a doubt that all humans—and by no means only members of their own community—deserved to be treated humanely. The Lord of the World himself demanded it: "But the stranger that dwelleth with you shall be unto you as one born among you, and thou shalt love him as thyself; for ye were strangers in the land of Egypt: I am the Lord your God."[9] God extends his mercy not just to the Children of Israel, but to all people, according to a psalm that was probably written after the Babylonian exile: "Yea, all kings shall fall down before him: all nations shall serve him. For he shall deliver the needy when he crieth; the poor also, and him that hath no helper."[10]

Not long thereafter, the first Greek philosophers founded Western ethics. The Sophist Lycophron declared that all people have by nature the same rights and that the state must protect the weak from the strong. His colleague Alcidamas expressed an even more daring idea: "God has left all men free; nature has made no man a slave." And Socrates made virtue the object of philosophical knowledge. According to him, convention cannot regulate human behavior, but rather the good emerges solely from knowledge of the world. More radically than its predecessors, this idea of Socrates explains why moral principles must be universally valid:

The search for right action is at the same time an approach to truth. And he followed this insight during his trial in the year 399 BCE, when he disproved all the accusations of his prosecutors, refused the escape plan organized by his friends, and drank the hemlock.

A morality that embraces all humanity was preached in India as well. Early calls for nonviolence show up in Vedic texts from about the eighth century BCE. After all, they say, a common soul connects all creatures. Since humans share the same essential being with animals and even with plants, people who harm others in the end are harming themselves. As the Isha Upanishad, one of the most sacred texts of Hinduism, declares, "When a man sees God in all beings and all beings in God, and also God dwelling in his own Soul, how can he hate any living thing?"[11]

In the early years of Hinduism, believers understood such a question metaphysically. Accordingly, it was forbidden to slaughter animals other than by following the correct ritual, while war was regarded as a normal phenomenon of human existence. But around the year 450 BCE, soon after the death of Confucius, both the Buddha and the North Indian prince Mahavira radicalized Hindu teaching. From the Vedic article of faith they derived an absolute prohibition against inflicting suffering on any living being, whether human or beast. The Buddha taught his followers to treat all beings with loving kindness and active sympathy at all times. Mahavira, still invoked today by the almost five million followers of Jainism, went even further. Jains should not only be peaceable and strictly vegetarian, they should also make every effort not even to harm an insect. Strict followers of Mahavira were not allowed to eat honey (because Jains regard robbing the bees of their food as violent), practice agriculture (because tiny animals die when fields are plowed), or even light a fire (because an insect could stray into the flames).

From Heaven to Earth

However differently Confucius, Mahavira, the Buddha, Socrates, and the authors of the Mosaic commandments justified their teachings, they all agree in their understanding of what actually constitutes religion. It is no longer the gods who are in the foreground, but humans' relations with one another. Rabbi Hillel, an important teacher in the time shortly after the birth of Christ, expressed this change in a nutshell. When asked to describe the essence of Judaism, the sage answered without a trace of otherworldliness, "What is hateful to you, do not do to your neighbor, that is the whole Torah, while the rest is commentary."[12] This requirement would go down in the history of philosophy as the Golden Rule.

Until about 500 BCE, every religion consisted in believing in the power of the gods and making sacrifices or performing other rituals in order to propitiate them. But now the sages of East and West declared the supernatural to be of secondary importance. The Buddha and Confucius refused even to speculate about God, and the authors of the Torah carefully avoided making any definitive statements about YHWH, the Creator, whose name no one must pronounce. Translated, Yahweh means only "I am who I am."

Myths were displaced by action in the present—a universally valid ethical norm. What the Roman politician and philosopher Cicero wrote about Socrates is equally true of Confucius, the Buddha, and the authors of the Torah: "Socrates was the first who brought down philosophy from the heavens, placed it in the cities, introduced it into families, and obliged it to examine into life and morals, and good and evil."[13] It was no longer a question of believing in ultimate truths. What counted instead was how one behaved toward others. If you overcame the boundaries of your own ego and those of your community by acting empathetically, you were following the new teachings.

It had always been characteristic of every religion to distinguish between believers and nonbelievers. After all, a shared faith reinforced altruism within the group.[14] Differences in skin color, language, wealth, and many other things can serve to divide communities from one another, but no division ran—and none today runs—as deep as the belief that one is in sole possession of revealed truth.

But precisely this division between "us" and "the others" was put into question by the philosophers of Judaism and Hinduism and the founders of Chinese ethics and Buddhism. They counseled taking the boundaries between communities less seriously than the things that bind all beings together. Five centuries later, precisely this leitmotif appears in the New Testament. Jesus, who said he came not "to destroy the law, or the prophets . . . but to fulfil,"[15] stood up for society's outcasts. He healed the lepers, saved the adulterous woman from being stoned, and even defended prostitutes and the hated Roman tax collectors: "Verily I say unto you, That the publicans and the harlots go into the kingdom of God before you."[16] This is the challenge in the Gospel of Matthew to his pious audience.

The Axis of Goodness

From a modern perspective, it is easy to understand why the sages of antiquity saw an all-encompassing love for one's fellow human beings as a spiritual act: Although we all are born with a predisposition to be selfless, evolution limits its initial reach to our own group. Thus if someone behaves altruistically to all fellow humans without distinctions—and possibly even to animals—it is a triumph of spirit over nature.

It is one of history's fascinating riddles why this new ethics arose at approximately the same time in such geographically dispersed locations as India, China, Greece, and the Near East. The Heidelberg philosopher Karl Jaspers called this spiritual turning-point the "axial age" because he regarded these cultures of the Near and Far East and the Mediterranean as points on an intellectual axis.[17]

It is notoriously difficult to compare the past of various peoples—too many chance happenings and unique events determine their histories. But we can discern some reasons for an intellectual turning point in these advanced cultures around 500 BCE. Cultural exchange probably played little or no role, since the physical distances were so great. Rather, there were some developments that ran parallel.[18] At the beginning of the first millennium BCE, iron-working had become established in India, the Mediterranean, and China; men now had access to tools of unprecedented effectiveness. Using heavy iron plows, peasants could make more land arable. Newly domesticated crops such as sesame, rice, millet, wheat, and fruit flourished in fields and orchards. With wealth on the increase, the division of labor also accelerated. More and more people could devote themselves to intellectual work as merchants, priests, and officials. They had the time to think and the education to challenge the old myths. Writing, a medium in which people could pass on their philosophical ideas, gained in importance.

Cities grew larger and became centers of trade. Thus they were the ideal breeding ground for norms that included strangers—distant trading partners, for instance. As the cities flourished, the ties to one's clan lost their importance. The growth of the state through wars and political marriages also played a part. Instead of living in small and easily comprehended villages and clans, people now lived in larger groupings and in close proximity to others to whom they were not related. Nevertheless, they had to learn to live in harmony. The rituals involving the Ark of the Covenant that are quoted in Leviticus were supposed to unify the tribes of Israel. Many of the ceremonies in the Chinese *Book of Rites*—ascribed to Confucius but probably older—had the same function.

And the larger the groups became, the stronger was the temptation to live at the cost of others. In a small-scale society of 150 members, everyone still knows what everyone else is doing. Freeloaders cannot remain undiscovered for long and are promptly punished. In the anonymity of a city, however, social control no longer suffices. Such a larger community needs a stronger ethic. Religiously founded norms could have fulfilled that function and served to discipline individuals.

Data from contemporary societies support this conjecture. For one thing, the larger the settlements are in which people live, the more value they seem to place on fair dealing. The anthropologist Joseph Henrich demonstrated this by having hunter-gatherers, villagers, and city-dwellers throughout the world play Ultimatum.[19] The urbanites were much more likely than the villagers to punish an anonymous miser at their own expense. And the villagers in turn used sanctions more often than did the hunter-gatherers. Paradoxically, the city-dwellers were acting much more strongly against their own self-interest. While it may be worthwhile to a villager to teach a lesson to a miscreant, the inhabitant of a large city is much less likely ever to meet up with the wrongdoer whom he punishes. It is thus obvious that in larger settlements, stronger norms for altruistic behavior exist.

For another thing, the more members a society has, the more likely they are to believe in a god who pays attention to their deeds, good or bad. That was the conclusion reached by a comparison of 186 different cultures on all continents.[20] Few hunter-gatherers believe in a just god who insures order among humans; they can regulate their affairs among themselves. On the other hand, the world religions all preach that justice is governed by higher principles. The bad are already punished on earth, and when they die they go to hell or are reincarnated as lower beings. The good are promised eternal bliss. And such promises are effective. No matter where they lived, participants in Henrich's Ultimatum experiment were consistently fairer when their society was influenced by one of the world religions.

A Feast for the Enemy

To be sure, it cannot have been just living together in larger and larger groups that led to the advance of morality around 500 BCE. Jaspers's "axial age," after all, was also an era of political turmoil. As the numbers of the rich grew, so did their armies, with soldiers now armed with iron weapons. To lose a war meant catastrophe. In Greece, the city-states, especially Sparta and Athens, struggled for hegemony and were later all threatened by the armies of the great Persian king Darius. In Northern India in the sixth century BCE, sixteen kingdoms existed whose kings ruled their subjects through military force and brutally put down the frequent uprisings against them. Ascetic movements of all sorts attracted followers; terrified by the cruelty of the world around them, more and more people sought peace within themselves.

In China, the Zhou Dynasty was crumbling. Regional rulers attempted to fill the power vacuum, but they kept getting entangled in civil wars. Confucius thought that only a new ethics could bring this self-destructive violence to a halt. And the Children of Israel were still suffering under their subjugation by Nebuchadnezzar II. The Babylonian ruler had destroyed the Temple in Jerusalem in 586 BCE and abducted almost the entire Jewish elite. In exile, the defeated Jews began asking themselves if Yahweh really was what they had thought him to be: a god of war fighting on the side of Israel. They also learned that reconciliation is often the best strategy.

And this was how the admonition to do good especially to one's enemies must have found its way into the scriptures a half-century before Christ: "If thine enemy be hungry, give him bread to eat; and if he be thirsty, give him water to drink: For thou shalt heap coals of fire upon his head, and the Lord shall reward thee."[21] This curious image does not refer to some perfidious torture, but to the way bronze was produced. As the heat melts the metal out of a rock, a good deed will make your enemy's hostility melt away.

The Old Testament Book of Kings, written during or shortly after the Babylonian Exile, relates an incident from Solomon's war against the Syrians.[22] Elisha, a prophet in the service of the king, lures the devious Syrians into his own trap in the mountains of Samaria, where King Solomon intends to slaughter the enemy troops. But Elisha counsels him to set a banquet before the Syrians instead, and then let them go. This ends the conflict: "So the bands of Syria came no more into the land of Israel."[23]

Who's Afraid of the Black Man?

As foresightful as Elisha's suggestion was, it must have been difficult to follow. For loving your enemy involves not only fighting down the thirst for revenge, but also being optimistic that the enemy will respond to your friendly advances—as if it were not already enough of a challenge to show generosity to strangers with whom we have *no* history of animosity!

The New York neuropsychologist Elizabeth Phelps has investigated the psychic forces that keep us from helping people outside our own community. She studied the unconscious prejudices that white Americans harbor toward their darker-skinned compatriots.[24] Her subjects were by no means intransigent or ignorant; almost all of them were urbanites with liberal convictions—or they thought they were, at any rate. Asked about their opinion of African-Americans, they gave positive or at least neutral responses.

But later tests of their subconscious reactions uncovered a much less flattering result. For example, Phelps showed her subjects pictures of faces alternating with words. The faces looked either African or European; the words designated either something good or something bad. The subjects were asked to push the appropriate one of two buttons. In the first round, the black faces and words like "love," "peace," and "beautiful" were assigned to one button; when white faces or words like "hate" and "ugly" appeared, subjects were supposed to push the other button. In a second round, it was the reverse. Now the black faces were paired with negative words, the white faces with positive words—and the task was accomplished much more quickly by the subjects. Despite their self-proclaimed enlightened attitudes, they obviously automatically associated their own skin color with "good" and darker skin with "bad." Phelps' results were even clearer when she performed CAT scans of her subjects' brains; when they saw a black face, a signal was immediately sent by their amygdala, a center in the brain that releases negative emotions. Our amygdala always reacts faster than our conscious mind. The latter can only suppress, as far as possible, the nascent feelings of fear and hate.

The experiment, however, does not prove that racism is innate. On the contrary, when Afro-Americans took the same test, their reactions were no different from those of their white compatriots. As soon as they saw black faces, their amygdala also sent signals of fear and hostility.[25] The deep prejudices of the white majority seem to have seeped into the heads of even African-Americans themselves. How powerful these acquired aversions can be is the real point of Phelps' results. Five decades of effort to overcome Americans' habit of sorting people by skin color have indeed changed—superficially—the way Americans think. It's no longer acceptable to make jokes about people who look different. But too often, feelings continue to be governed by what group someone belongs to.

A Change of Perspective

It is all the more astonishing, then, when someone like Otto Weidt treats others without regard to where they come from, but simply as fellow human beings. How can such universal goodwill be brought about? The ancient sages agreed that it is of little help to preach morality. First, because all-encompassing love of one's fellow humans must be practiced, and second, because people have to get used to the corresponding change in orientation within themselves.

Every teacher recommended his own method for accomplishing this change. Confucius hoped that his strict, ritualized rules of behavior would induce people to put their own egos second. The Jewish tradition places great value in the study of scripture. Greek philosophers emphasize the search for truth in learned discourse. Buddhism recommends cultivated sympathy for all beings through meditation; its practitioners should first try to discern friendly feelings for themselves and people close to them, later to include people who are indifferent or even hostile to them, until finally their goodwill encompasses the entire universe. More ascetic paths such as the several varieties of yoga attempt to guide their adherents into a dimension of being that is more significant than the world of their private needs. Christianity recommends that believers always keep sight of the boundless devotion of Jesus.

As much as these practices differ in their details, there is one thing they all urge: Humans should learn to change their perspective. That is how our imagination creates the precondition for caring for other people without regard to their background. Whoever can really put himself in the shoes of another no longer asks what group that person belongs to. Thus morality exploits our innate ability to be empathetic, raising it to a higher and more conscious level.

And the expressions of the Golden Rule, which occupies such a central place in all these teachings, resemble one another down to their choice of words. Once more, Confucius may have been the first to express it: "Do not impose on others what you yourself would not desire."[26] "Hurt not others in ways that you yourself would find hurtful" says the Buddha.[27] "Do not do to other men what would outrage you if you had to experience it yourself," is the way a speech of the Sophist Isocrates puts it. At the end of the Sermon on the Mount, Jesus refers to the Golden Rule, as does the Indian epic *Mahabharata* in its "Book of Instructions," the most important compendium of Hindu ethics.[28] And the medieval Syrian scholar An-Nawawi reminds his fellow Muslims that the Golden Rule is as valid toward nonbelievers as it is toward their co-religionists.[29] Perhaps the most moving of all the formulations of the Golden Rule is that of Martin Buber in his German translation of the Old Testament: "*Liebe Deinen Nächsten, denn er ist wie du*"— "Love your neighbor, for he is like you." No further explanation is needed: Everyone understands immediately the significance of the four simple words "he is like you." Instead of warning and moralizing, it's enough to remind people of a simple truth: Measured against all we have in common, the differences between us are tiny. Whatever another person does, whatever is going on inside her, it is not foreign to us. We have already felt the same thing ourselves.

Both the rationalism of Greek philosophy and the doctrines of all the world religions place value on self-knowledge. It is a decisive step toward acting according to the precept of comprehensive love of one's fellow humans, and the individual achieves a new significance thereby. People who understood themselves as part of a group became individuals who knew they were unique and yet extremely similar to one another. Loyalty to the community, which can include such dark concomitant phenomena as xenophobia, is reduced in significance, to be replaced by the personal relationship to the other.

The story of the Good Samaritan is about such a change in perspective. Jesus tells the parable in the Gospel of Luke in answer to the question: Who is the neighbor I am supposed to love as myself? The message to his listeners was clear: The priests who pass the wounded robbery victim without taking notice of him may have belonged to the elite, but the Samaritan, the member of a community despised by the Jews, is vastly superior to them in the morality of his actions. Yet Jesus sharpens the message further; he asks in return who *the victim's* neighbor is. The man who took pity on him, responds his questioner. For now it is the victim who feels himself bound in gratitude not to his own people, but to the helper. By selflessly helping a stranger, the Samaritan overcomes not only his own group loyalty, but also the group loyalty of the victim.

The Limits of Loving Thy Neighbor

The Golden Rule is not an invitation to exploit one another. Although it is often confused with the principle of "Be good to others so they will be good to you," it is not concerned with such reciprocal altruism. Its precept is to "do unto others, *as you would have them* do unto you." How others actually treat us doesn't enter into it. Because it is valid without exception, the Golden Rule is a norm. It is the oldest known commandment that requires a certain behavior not just to members of one's own community, but to all humanity.

But is such a precept compatible with Darwin's ideas? Even if one assumes that evolution operates on several levels—genes, individuals, and entire groups compete with one another—there is no mechanism that favors selflessness *toward everyone*. For as the previous chapter showed, altruism is successful when the competition is shifted to the next-highest level, for then those who attach themselves to a group have the advantage. It follows that evolution can never produce selflessness toward the entire species, because that is the highest level.

History seems to support these objections. Neither Confucius nor Mengzi, the authors of the Torah nor Socrates, Jesus nor the Buddha reflected the norms according to which their contemporaries lived; they were all more or less solitary voices. Nor has any world religion ever lived up to its ideals. Christianity produced the Crusades and the atrocities of its missionaries; Islam produced holy wars; Hinduism produced pillaging, murdering mobs. Even Buddhism, with its reputation as especially peace-loving, repeatedly nourished violence—for example, the forced conversion of the Mongols in the sixteenth century or today's struggle between the Singhalese and the Tamils on Sri Lanka. Not even the priests, teachers, and monks whose task it is to lead believers on the path to universal love lived (or live today) as the sacred texts of all the great religions prescribe.

On the other hand, however, the glass is also half-full. We are very far removed from a world in which people do something for others *only* if they belong to the same community—even though our innate altruistic inclinations alone cannot explain the amount of humane helpfulness.

What's more, the norm of treating people well regardless of their background has long since freed itself from its religious roots. Probably the most famous further development of the Golden Rule is the categorical imperative. With that principle, the eighteenth-century German philosopher Immanuel Kant overcame two justifiable objections. The first was that the Golden Rule is usually negatively formulated; it says what we should *not* do, not what we *should* do. The second is that the change in perspective of an individual is often not sufficient to make a moral judgment. If all decisions had to depend on what individuals would wish or not wish for themselves, no judge would be able to put a murderer in jail and a masochist would find it right to torture other people. So Kant declared that instead of one's own personal standpoint, one should adopt the impersonal standpoint of a legislator: "Act so that the maxim of thy will can always at the same time hold good as a principle of universal legislation."[30]

Rome: A Model for Success

Thus universal norms have not merely survived for at least two and a half millennia, but have also spread and continued to evolve. The riddle of how that happened finds an answer in the complexity of our society. Among the ants or the apes, each group looks after itself. But the high cultures of antiquity already had various communities living together and depending on one another. The Jewish elite in Babylonian exile, for example, did not sit enslaved and weeping on the riverbanks, despite what it says in Psalm 137.[31] They were allowed to trade, own real property, and even keep slaves. Jews also became high officials in Nebuchadnezzar's state. Thus the Jews lived, however unwillingly, as one community among many in the mixed population of Babylon and maintained lively intercourse with the other groups.

The Greek city-states, too, were not insular tribal communities; rather, they were comprised of smaller groupings with very different rights, as described in Chapter 8. Social intercourse between people of different backgrounds then reached a high point in the Roman Empire. The Roman historian Livy, writing at the time of Christ's birth, declared that from the very beginning, the Romans had been a conglomeration of immigrants. In the late empire, people everywhere in the known world were trading with one another, and it was common for families to move from one shore of the Mediterranean to the other or even into chilly Germania. They were able to retain their cultural identity in their new surroundings—Rome allowed everyone his or her own religion.

Justice and morality, declared the philosopher and politician Cicero, could only be founded on the basis of solidarity with one's fellow humans and affection for the whole human race. Humans, he wrote, are endowed with divine reason, and thus right and wrong could not depend on language, class, birth, or religion.

To some extent, the Roman Empire even corresponded to Cicero's ideal of universal justice. At least it came nearer to it than any other ancient culture did. For at all times, Rome was a remarkably open society. Already by the first century BCE, the republic declared the male inhabitants of all the cities south of the Po River the equal of native Romans. Later, entire cities in the colonies were granted the rights of Roman citizens. Foreigners and even manumitted slaves could gain citizenship.

Of course the state profited from the loyalty of its new citizens. Neither the economy nor the army of the gigantic empire would have been so successful if its rules had not united as many people as possible. Social diversity and globalization thus favored the rise of universal norms.

The Treasure of Reputation

A morality that includes all humans benefits not just the community but the individual as well. In a small group, everyone can easily observe whether a fellow member is behaving decently, but people in larger societies usually must gather their information indirectly—by gossip. As unpleasant as gossip can be, it enables a new form of selflessness: To help a stranger can pay off if others witness the deed and from then on consider us trustworthy.[32]

"The purest treasure mortal times afford, is spotless reputation," declares the accused Mowbray, Duke of Norfolk, in Shakespeare's *Richard II*.[33] Anyone who has done business on eBay knows how valuable others' opinions of us are. People don't just prefer buying things from complete strangers who have acted fairly in earlier transactions and thereby earned many stars, they are also willing to pay them higher prices than they would pay to sellers without the same reputation. In experiments, subjects prefer to help people whom they regard as reputable, even when those people have never directly done anything for the helper.[34] And participants in economics games are willing to turn more money over to a partner if they know that person has previously donated money to good causes.[35]

It makes no difference whom the other person has helped—provided it is not a well-known evildoer.[36] In the just described experiment, the money went to the Third World; on eBay too, as a rule we don't know who evaluated the seller.

Reputation is a key to understanding how selflessness can survive beyond the border of a group. It operates more subtly than competition between groups. Both mechanisms prevent altruists from being exploited and eventually dying out. But while group selection demands that the gift of the altruist benefit the community, reputation, a more recent evolutionary phenomenon, does not depend on the goods the altruist gives. Now, voluntary sacrifice serves as a signal that a person's intentions are sincere. Altruists can be recognized thereby; their advantage is now that people are willing to trust them.

Every action that says something about one's personality can serve as such a signal, no matter who benefits from it. That's why the pursuit of a good reputation favors the kind of selflessness that ignores the borders of one's own community. It only requires a consensus about what a society regards as a good deed and thus a sign of trustworthiness—in other words, a norm. For example, the general conviction could become established that someone's probity is proved by their dealings with their neighbors.

The only problem is that such narrowly defined norms open the door to swindlers. A clever rogue will be obliging to his neighbors and use his reputation to rob everyone else. More comprehensive precepts reduce this risk. And the more complex a society is, the more it needs norms that are valid without exception, like the religious commandment to love one's neighbor or Kant's categorical imperative. For if we draw the circle of people we treat with fairness and generosity too narrowly, our credible reputation is limited to that small number. On the other hand, it costs more to be fair and helpful to everyone. Is that why the Golden Rule began to spread only about 2,500 years ago? Obviously it can only survive in a relatively highly developed society. The spread of Islam in Africa since the eighth century BCE illustrates how the transition to a world religion with strong norms allowed trade to flourish; people who wanted to benefit from the new opportunities had to join the new religion.[37]

Every norm develops because it promises an advantage. But once it becomes established, it leads a life of its own. For as Chapter 8 showed, people behave according to what is customary, even if in individual cases it does not benefit them. That's why we tip generously when we're on vacation even if we'll never see those waitresses and waiters again. And when well-to-do Europeans and Americans donate money for the care of children in the developing world, usually no one except the tax collector knows anything about it. The donors are not giving because they want to enhance their reputation, but because they have internalized the norm that everyone has a right to education and medical care.

CHAPTER 11

The Triumph of Selflessness

IN 1980, RICHARD STALLMAN ASKED A CO-WORKER for the program code for his printer. He was twenty-seven years old and in love with computers, and he'd broken off his study of physics to work as a humble programmer at the Massachusetts Institute of Technology. He was annoyed that his printer didn't let him know when it had run out of paper or alert him to a paper jam. He wanted to improve the program. But the program's author refused his request: His boss had made him promise not to release the code.

This refusal was the beginning of a story that would prove how much the balance of power between egocentrics and altruists has shifted in recent decades. His co-worker's secrecy made Stallman furious, he recalls; that wasn't how hackers were supposed to treat each other. At the time, the word "hacker" did not yet smack of computer crime but instead was what young—mostly male—computer enthusiasts called each other. A hacker was someone who really understood how computers worked, spent the night in front of his screen out of the pure pleasure of writing elegant programs, and shared the fun with others.

But lately, more and more hackers had been breaking the unwritten rules. Firms like the one still spelling its name "Micro-Soft" took possession of what had belonged to everyone and kept its program codes under lock and key. A stir was caused by an open letter to the hacker community declaring that software was property and using it without paying for it was theft. The letter was signed "Bill Gates."

Stallman was not willing to accept the new convention. He believed that technology should serve humankind and be as good as it was possible to make it. To declare a program private property was in his eyes an enormous waste of human effort. It rankled him deeply that he was unable to improve the software for his printer because he was denied access to the code. On September 27, 1983, Richard Stallman threw down the gauntlet via the computer newsgroup net.unix-wizards: "Starting this Thanksgiving I am going to write a complete Unix-compatible software system called GNU (for Gnu's Not Unix), and give it away free to everyone who can use it. Contributions of time, money, programs and equipment are greatly needed."[1]

At the beginning of 1984, Stallman quit his job and from then on earned his money as a freelance software consultant. Later he was paid by a foundation. First, he programmed an editor that, somewhat similar to Word, serves to enter text into the computer. Then he authored a compiler that translated the program instructions legible to humans into the language of the machine, consisting of ones and zeros, into program instructions legible to humans.

The lone wolf from Boston quickly gained enormous respect in computer circles. What Stallman gave away for free was as good as the expensive commercial programs, and in some cases far better. (The GNU editor and the compiler are still being used today. In the computer world of short-lived programs, hardly anyone even remembers most programs from the 1980s, but Stallman's creations and their ongoing elaborations continue to be standards.) Anyone who wanted the code for the software could have it. To prevent profiteers from taking over his work, Stallman had one condition which in retrospect he calls his best hack ever: Anyone may use, change, or expand the GNU code as they wish—but they must make all the programs they develop freely available in their turn.

By the end of the 1980s, however, Stallman had begun to reach the limits of what he could do alone. But then he got unexpected help from far-off Helsinki. Linus Torvalds, a twenty-one-year-old computer science student, was working on the central program that Stallman lacked, one that could oversee all processes on the computer. But whereas Stallman relied on his own capabilities, Torvalds made use of a new method to bring other developers into the process: the Internet. By 1993, more than 100 developers all over the world were working for free on the operating system now known as Linux.

That number has grown to tens of thousands today. It is common knowledge that Linux has been an unprecedented triumph. By now, not only do millions of desktop computers run on the system: It is also employed by the nerve centers of the information society—more than half of all the servers that feed data into the Internet. Anyone who has done a search on Google has been connected to a Linux computer.

Other software developers were also inspired by Stallman and began to make their work available to everyone for free. "Without free software, the Internet in its present form would be unthinkable," says the technology writer Christian Imhorst.[2] The principle is so far superior to the tradition of programming behind closed doors that for the time being, even for-profit companies publish their in-house applications for free. IBM, for example, supports the continued development of Linux with 100 million dollars annually in order not to be dependent on Bill Gates' Windows. Most of that contribution is made by developing, at its own expense, components for Linux and then making them available free of charge.[3] And Google has put into circulation a complete software system for cell phones under the GNU constraints. The Microsoft executive Steve Ballmer angrily compared the programs licensed by Stallman to a cancer; the freedoms that the annoying Boston hacker had built into his programs would infect everything that came into contact with the descendants of his software.

In the end, Stallman's habit of giving away his knowledge began to prevail outside the world of computer science as well. Many more than one hundred thousand volunteers on all continents write for the Internet encyclopedia Wikipedia, which aspires to collect the knowledge of all humankind. A good 15 million articles in more than 200 languages have already been written for it. So that everyone who writes or edits an article can be sure that their contribution will always be available free of charge to everyone, the entire project adheres to the principles of GNU.

But even the mighty Wikipedia is only a tiny part of the selflessness on the Internet. Who among us hasn't given advice to some stranger seeking help from an online forum? By now, so many people regularly share information on the Internet that no one tries to keep count of them anymore. The last estimate from 2008 counted 133 million blogs on which people share their ideas and knowledge. Very few of them earn money with ads on their websites. Direct benefits to the authors in general are very limited—they write, take pictures, film, and publish their works for the joy of sharing.[4]

The Global Tire

When the indomitable Stallman began developing his free software on his own in Boston, no one could have imagined that only two decades later, hundreds of thousands of people around the world would work together without pay to create communal products like Wikipedia and Linux. In the meantime, such forms of cooperation are spreading so quickly that we can hardly foresee where this development will lead. Perhaps future historians will someday compare it to the transition to communal hunting in the early Stone Age, when our forebears realized how much more reliably they could feed themselves if each family no longer had to fight for survival on its own. Thus arose tribal groups and norms, and the willingness to share became established.

In an analogous way today, we need paths by which we can shape communal life on a global scale. The problems of the earth obviously can no longer be solved by using the rules and institutions that were more or less effective on the smaller scale of families, settlements, regions, and countries.

Within a very brief period, people on all continents have become dependent on one another. Today, hundreds of thousands of men and women of the most varied backgrounds work together not just on collective enterprises on the Internet, but also in transnational corporations. One can see the extent of the change clearly by comparing the means of transportation used by our great-grandparents with our own. A hundred and fifty years ago, the breeder from the next village provided horses; you either built your own wagon or asked your neighbor the carpenter to help, and you fetched the wood you needed from the forest. On the other hand, modern cars, trains, and airplanes consist of thousands of parts brought together from all corners of the globe. Even the materials for an ordinary automobile tire come from a dozen different countries: the rubber from Vietnam and Malaysia, the polyamide fiber weave from China, and the steel belt from Sweden.

How dramatically the interrelation of the continents is accelerating can be measured by the so-called trade ratio. It indicates what percentage of a country's total economic output is made up of imports, exports, and services. In Germany today, it is more than 70 percent. Worldwide, the percentage of goods that cross national borders has almost doubled since 1990 and more than tripled since 1950. The rate of increase is even greater when one includes services performed for foreign countries.[5]

While more and more people work with—and have constant contact with—geographically distant colleagues, traditional ties are loosening. In the postwar generation, it was still typical to feel closely tied to just a few groups. One's identity was encompassed by being the citizen of a particular place, being a Catholic or a Democrat. For employees who proudly identified themselves as "Ford workers" or "US Steel workers," it was a matter of course to stick with their factory through good times and bad. People knew what they could expect from their communities and were ready to give something in return. But in an era of sharper global competition, companies seldom promise secure, lasting jobs— and the price, of course, is that they can have less confidence in the productive engagement of their workers. Many highly qualified workers prefer to gain experience in New York today and in Hong Kong tomorrow. However, where people must seize opportunities where they can, they are all fighting for themselves alone.

The Bankruptcy of Nations

One of the greatest illusions of the last few decades was that an economy founded on the pure self-interest of its participants would reliably provide for the needs of humankind and achieve optimal results. That illusion reaches back more than 200 years to the astute but often misunderstood analysis of Adam Smith. At the time when the first factories were being built, the question was how the new opportunities opened up by industrialization should best be exploited, and how to get hundreds—and later, thousands—of people to create something together.

With his answer to that question, Adam Smith, a professor at Glasgow University and later an independent scholar, became the father of modern economics. With the production of pins as his model, he explained how the increasing division of labor was leading to an enormous increase in production. A single craftsman without machinery could "scarce, perhaps, with his utmost industry, make one pin in a day, and certainly could not make twenty."[6] Ten workers in a specialized factory, on the other hand, could turn out 48,000 pins a day. But for the division of labor to work, people had to exchange things. And in this exchange, everyone was looking for their own profit: "It is not from the benevolence of the butcher, the brewer, or the baker, that we expect our dinner, but from their regard to their own interest. We address ourselves, not to their humanity but to their self-love, and never talk to them of our own necessities but of their advantages."

Nevertheless, the egocentric efforts of all citizens were supposed to lead in the aggregate to general welfare. Even if every individual "intends only his own gain," the market would regulate itself. Every merchant would be "led by an invisible hand to promote an end which was not part of his intention."[7] The laws of supply and demand would ensure the optimal distribution of goods. Where the need was greatest, the highest profits were to be expected, and since any smart merchant would invest his money there, he would meet people's needs in the best possible way. This was the position Smith argued in *The Wealth of*

Nations, first published in 1776. Its principles are still invoked today by economists and politicians.

The theory that Smith propounded and that his numerous followers continue to defend is well founded. An economy driven by pure ego-centrics would indeed function—but only as long as all human interaction were amenable to regulation by the market. Thus except for their transaction, a seller and a buyer would have nothing to do with each other. One of them gets the merchandise; the other gets the money (or other merchandise in exchange). And third parties who are not participants in the business would not be affected by it.

One can argue about whether this conception was unrealistic even in the eighteenth century. In the view of the economic historian Joel Mokyr, the Industrial Revolution originated in Great Britain because English businessmen followed a strict moral code and trusted one another.[8] But that also means that business deals have a decisive influence on uninvolved third parties, an influence hardly expressible in terms of money: Whenever two partners treat each other fairly, they strengthen the trust others place in the norms—and thereby also others' willingness to risk entering into deals themselves.

Today, at any rate, such effects are omnipresent, for in a highly networked world, every action casts unimaginably wide ripples. The financial crisis of 2008 offers a perfect example. When the New York bank Lehman Brothers became insolvent in September, only its immediate creditors should have been affected, according to classic economic theory. But in fact, when the collapse of Lehman Brothers became known, the entire system of monetary transactions between banks all over the world broke down completely. Everyone feared a chain reaction. Even if a borrower had never done business with Lehman Brothers, he had reason to fear becoming insolvent, especially if a third bank owed money to him and had also lent money to Lehman Brothers or to one of their business partners. Moreover, it turned out that the once highly regarded Lehman Brothers had concealed its risky investments before it went bankrupt. So how could anyone be sure that other banks weren't also playing with marked cards?

No one trusted anyone anymore, with the well-known result that the banks stopped renewing the credit that industry needs in order to survive. And soon thereafter, people in America and then in other places began to keep their wallets in their pockets.

During the entire crisis, all participants acted completely rationally from their own points of view, just as Adam Smith once described and recommended. However, far from increasing the wealth of nations, according to estimates of the International Monetary Fund, the financial crisis decreased it by an incredible 12 trillion dollars ($12,000,000,000,000).

In order to function, markets need a basis, and although they cannot produce it by themselves, they can certainly destroy it. Trust is such a basis.

Capital and Cod

An economy that intends to promote the general welfare simply by channeling egocentrism is an open invitation to freeloaders. Indeed, parasitism was the deeper cause of the 2008 financial crisis. In the preceding years, private investors, banks, and even entire states had all tacitly abandoned the norm that one should not live above one's means. Those who spent more than they earned were no longer punished, but encouraged. And everyone profited from this shift of values: Citizens lived in houses they actually couldn't afford. Banks did a land-office business in sub-prime mortgages whose risks they passed along to other investors. Politicians made voters happy with generous benefits willingly financed by creditors in other countries. This system could continue to function only as long as all of the players trusted one another. Everyone was juggling in their own ways with risky credit in the hope that they could pull out of the game at the right time—and thus leeching mutual trust out of the economy.

It's not just the financial crisis that shows what enormous harm can be done to our highly networked world by freeloaders. When a fishing fleet left Glasgow harbor in Adam Smith's time, there was no reason to fear that it would perceptibly decimate cod stocks in the North Sea. The ocean provided more than enough replacements. Today, thanks to more and more aggressive fishing techniques, so many animals have been turned into fish filets that cod stocks in many places have collapsed. The operators of the first steam-driven machines at the beginning of the Industrial Revolution didn't have to worry about emitting coal smoke into the air. When an energy company does so today, it contributes to global warming and the threat that entire countries like Bangladesh will be inundated by rising sea levels.

Adam Smith was very well aware of the basic problem that the market cannot provide communal property because no one wants to pay for it. His view was that in such cases, government must intervene. The ruler of every land must ensure secure borders and provide reliable courts, roads, dikes, and education for his subjects.

But in a globalized world, governments increasingly fail to establish these basic necessities, for no one has hegemony over the air and the oceans. Worse still, when a country takes more than its share of these resources—for example, by emitting above-average amounts of greenhouse gases—everyone in the world suffers the consequences. Effective treaties to protect the atmosphere or the cod have failed because of this kind of freeloading.

And it seems no easier to tame the international financial markets. No one disputes that a set of strict regulations could prevent future crises. But capital is more mobile than ever before, and a country that insists on keeping its lax standards can hope to attract money from all over the world. Thus we are all stuck in the prisoner's dilemma.

The Spoilers from Athens

Obviously, Adam Smith's economic recipes are of only limited application in today's world. On the other hand, societies leave a decisive resource untapped when they count on markets and the government alone to regulate citizens' lives. After all, there have been large-scale communities for at least one hundred thousand years, but laws and markets of significant scope have existed for only a few thousand years. For most of human history, our forebears cooperated with one another because a certain amount of selfless inclination was innate, and also because they hewed to group norms. This is the foundation on which our life together as humans is based, not on egocentrism channeled in free markets.

Humans' ability to be selfless developed in direct proportion to their increasing dependence on one another. From the communal concern for progeny and food in the early Stone Age to the admonition to love one's fellow humans that emerged in the world religions about 500 BCE, the history of humankind has been a history of stronger and stronger altruistic norms.

And it is clear that this tendency continues. It is not only that, as described in the previous chapter, the larger the community, the more often its members are willing to pay more for fairness. A comparison of modern societies also shows that a complex economy encourages more and more civilized rules for living together. Three central factors encourage this progress: The societies become more diverse, they become networked, and knowledge plays a larger role in them.

The Nottingham economist Benedikt Herrmann and his colleagues investigated how people in various world metropolises behaved in the Free Rider Game. In cities in modern industrialized countries—Boston, Copenhagen, Bonn, Seoul, and even in Chengdu in China—participants paid relatively large amounts into the common pot.[9] In cities in less developed countries, on the other hand—Istanbul, Riyadh, and also in Minsk—voluntary contributions were significantly lower. But it was the citizens of Athens who displayed the least communal spirit, although the experiments took place long before the Greek financial crisis.

Was the norm of cooperating with strangers weaker in the less-industrialized countries? When players in these countries are permitted to punish free riders, at any rate, they hesitated to make use of that opportunity. And when someone did decide to punish a free rider, strange things happened in big cities like Minsk, Istanbul, and especially in Athens: Dodgers who had to accept punishment did not think for a moment of changing their behavior and ponying up in the next round. Instead, they chose to invest in punishing some other player who hadn't freeloaded. Possibly they suspected that particular fair player of giving them the punishment and were now taking revenge. Of course, they couldn't be certain, because the punishers are always anonymous. At any rate, the free riders paid astonishing amounts to get satisfaction—and in so doing, completely destroyed cooperation in the group.

In cities like Boston and Copenhagen, by contrast, the parasites were quickly punished and accepted the penalty without complaint, as if understanding that they had deserved it. Surveys confirm these results. In the experiment, the more strongly citizens of their city believe in justice and the laws, the more reluctant the punished players were to take revenge. In these societies, people also find it less acceptable to cheat on taxes, keep a lost wallet, and disobey traffic laws—behaviors that people in Istanbul or Athens accept with a shrug.

Of course, inhabitants of these cities are not worse human beings, nor are they necessarily bigger egocentrics. And their behavior is understandable. It would be illogical to do much for strangers if they are likely to only be pursuing their own advantage. People in such societies feel more commitment to others within their immediate surroundings than to the general welfare. Paradoxically, wherever citizens display little communal spirit, their culture is strongly collectivist: The family is put first, the individual counts less—and of course, a more abstract entity like the state has an even smaller value.

Morality from Below

Conversely, it turns out that individualism does not promote parasitism. The experiments of Herrmann and his colleagues show exactly the opposite. Whether in Germany, Denmark, Switzerland, or the United States—in all societies that value the free development of the personality, people made large contributions to the common pot.

What seems to be a paradox at first blush is easy to explain. In an individualistic society, people belong to many communities at the same time, and these circles are not concentric on the model of "family—clan—village." Instead, they intersect. In the more traditional world of our grandparents, a village woman may have been a member of the church, belonged to the dairy cooperative, and been active in the sports club, but in reality she belonged to just *one* community, for the people she met in those various groups were essentially all the same people.

An urban woman of today, on the other hand, encounters a completely different group in her yoga class than in her church choir. Her strongest ally in the parent-teacher association may be a man who in his job is doing everything possible to weaken the company she works for. And many of her friends live in other towns—some even on different continents. Thus this woman comes to understand through experiences in extremely varied constellations that cooperation is worthwhile. Compared to her grandmother, she has a much larger choice of groups she can join. If she doesn't like the people in her yoga class, she can look for another one or switch to tai chi. The various communities compete for members and, as we have seen in Chapters 7 and 8, that promotes altruistic norms.

A comparative study from Italy that has become a classic in the field shows how a rich web of overlapping social contacts strengthens moral behavior and mutual trust.[10] In the late 1980s, the Harvard political scientist Robert D. Putnam set out to discover why some regions in Italy were models of good government while others were rife with corruption and mismanagement. He was able to trace the differences back to people's willingness to act toward the achievement of common goals. In areas like Emilia-Romagna in the north, for example, the social life of so-called civil society was flourishing. Although the neighborhood committees, athletic teams, and Rotary Clubs were for the most part apolitical, they strengthened the norms of the community. People were used to sticking to the rules and cooperating, and they made sure that others did, too. Thus politicians could not stay in power if they abused their position for personal advantage.[11] In regions where civil society was less robust, on the other hand, such figures had an easier time of it, since the norms were more relaxed. Later studies comparing European countries and American states reached the same conclusion: Morality and trust permeate a society from below.[12]

The Global Village Lives

If the willingness to act selflessly is increased by a multiplicity of group allegiances, then the networking of the world should have the effect that more and more people behave altruistically. There is much evidence to support that assumption.

The outpouring of global charity following the tsunami of 2005 and the earthquake in Haiti in 2010 would have been hard to imagine two decades earlier. Americans donated 1.8 billion dollars for the victims of the tsunami in the Indian Ocean alone.[13] And Germans gave almost nine hundred million dollars, a record donation for the country for disaster relief. This generosity could hardly be due to the television images alone, however heart-rending, for independent of current news reports, during the 1990s Germans donated increasing amounts for emergency and development aid.[14] Obviously, we feel increasingly responsible for the fate of people in poorer countries.

Of course, modern communications technology makes it easier to help faraway strangers. If you want to donate, you can do it with no trouble via text message; social media seem made for organizing support for worthy causes. When the invitation to give comes from a friend, you probably feel some pressure to join in the campaign, and you can trust that the cause is not a scam. With just such a snowballing campaign in 2008 on Facebook, in just a few weeks the young Colombian Oscar Morales not only succeeded in getting more than four million of his own countrymen onto the streets to march in protest against the FARC guerrillas, but also in sparking demonstrations against the South American hostage-takers in dozens of cities around the world.[15]

Besides these practical advantages, global networking above all changes our perceptions. When we leave a comment on a California website, have a chat via Skype with old friends in Shanghai or Cape Town, and spend our vacation on a beach in Thailand, the difference between near and far soon starts to blur.

The feeling of having the whole world as a neighbor increases peoples' willingness to work together. The American economist Nancy

Buchan was able to show this effect in six cities on five different continents.[16] She translated the metaphor of the global village into a game. Participants had the choice of paying money into a city pot (with four players in the group) or a world pot (with twelve players). The city pot took contributions from citizens of the players' own town, the world pot from people on other continents. And then, as usual in the Free Rider Game, the total in the city pot was doubled, the total in the world pot was tripled, and the new amounts were distributed evenly among the players. To level the playing field for participants from poorer countries, Buchan began the game by giving each person ten coins of play money. After the game was over, players could exchange their chips for the currency of their own country.

Obviously, it was potentially more rewarding but also riskier to contribute to the world pot. If other players didn't cooperate, there was a risk of loss. In both cases, however, paying in was a selfless act, for as we have seen in Chapter 7, the biggest winner is the parasite who pays in nothing but profits from the contributions of others.

The question was, would players be more generous with their compatriots or with the inhabitants of distant continents? Regardless of where they lived—in Milan, Tehran, Johannesburg, or Buenos Aires—participants paid an average of two of their ten coins into the city pot, but their contributions to the world pot varied widely. Citizens of Tehran invested two of their coins in the global as well as the city pot. But players from Columbus, Ohio, risked more than twice as much globally. They seemed to have more confidence in the fairness of people in other parts of the world. All the other cities lay between those two extremes: After Teheran came Johannesburg, then Buenos Aires, Kazan in Russia, Milan, and Columbus.

How can this sequence be explained? Buchan investigated how much the different countries had been affected by globalization. To measure its effects, economists have invented a "globalization index," a number that increases the more a country trades with the outside world, is visited by foreign tourists, accepts immigrants, or watches foreign films. And the amount players were willing to pay into the world pot rose in parallel with this globalization index. Thus the more open a society is, the more comprehensive is its concept of fairness and solidarity.

Differences between players from the same city were also dependent on their intellectual and social horizons. Those who listened to music from other countries, made international calls, had access to the Internet, or could speak a foreign language almost always paid more into the world pot than their less cosmopolitan compatriots. Even regular visits to restaurants specializing in foreign cuisines or the purchase of international products like Coca-Cola and Levi's jeans had positive effects. Obviously, consumers had internalized the fact that they were enjoying the benefits of cooperation across cultural barriers.

The way contributions were distributed between the city pot and the world pot is an optimistic sign: Participants who invested more in the global pot were not by any means more stingy toward inhabitants of their own town. Instead, the cross-border payments flowed in addition to the contributions to the local pot, which remained more or less the same. Thus globalization and worldwide networking do not merely cause a shift in our interest for other people. Instead, they release additional generosity. Seen in this light, altruism is not a finite commodity.

The Buffaloes of the Information Age

When early humans began to hunt together, they not only became more dependent on one another but also put their economy on a different footing. Here too, the analogy to our present situation is clear. We are in the midst of an economic revolution that is—in addition to social diversity and interconnectedness—the third factor motivating us to act altruistically.

Within just a few decades, the ways we earn our livings have radically changed. Fewer people work with their hands, more and more work with their heads. They invent, direct, and organize. Already today, German companies earn more than half their revenue with an insubstantial means of production: the knowledge stored in their employees' brains.[17] Even with a product as solid as a high-performance automobile, a third of its value resides in its electronic controls. And the programming of all the computers concealed in every car is increasingly important. Leading developers in the automobile industry expect that 90 percent of future progress will be in the area of electronics and software.[18]

The triumphant advance of the resource called information became undeniable in 2006, when Google became the most valuable brand in the world. The company has maintained that position every year since against firms from the traditional economy like Coca-Cola and McDonald's.[19] Thus the number one company is one whose customers pay not one cent for its multiple services. People who google are trading information for information. They reveal their personal preferences and get almost everything the Internet has to offer in exchange.

Google based its business model on the fact that the trade in information is governed by different rules than those that apply to physical products and services. A car, a house, even an oil field can be owned and thus easily exchanged for money or other possessions. A barber naturally only plies his comb and scissors for a fee, because every haircut costs him more work. Under such circumstances, goods can be rationally distributed according to the laws that Adam Smith formulated.

But whoever has access to knowledge can give it away without losing it. There is good reason to call an economy whose most valuable commodities are not things, but information, a "weightless economy," because knowledge is insubstantial. Since it's so easy to pass on information, in the long run it's almost impossible to maintain sole ownership of it. Restricting access to the fruits of your knowledge to paying customers requires a huge effort to catch and punish the ubiquitous pirates in the Internet. Knowledge producers face the same problem as a troupe of Stone Age hunters, according to the American anthropologist Samuel Bowles. They had to work together to bring down a buffalo, and we have to work together to program system software or unlock the human genome. Once the task has been accomplished, however, it does you no good to try to keep its fruits to yourself. A buffalo supplied more meat than the hunters could eat by themselves.

Thus the insubstantiality of knowledge encourages a culture of sharing. Science had transformed the world in the past 200 years. Its success has always rested on this principle: If individual scientists want their work to be recognized and acknowledged, they must publish their findings and allow them to be critically examined and also further elaborated. In almost every respect, a community whose members can feel free to help themselves to buffalo meat or research results will prosper more than one that sets up elaborate barriers. In the weightless economy of the future, it will be essential to share selflessly.

Humanity today stands before the enormous challenge of insuring cooperation on a global scale. Solutions that only apply to the interests of individual persons, companies, or countries will fail. We don't have much time left. The price of failure would be catastrophic developments such as unchecked global warming, unmanageable streams of refugees, and endless wars over resources.

And yet, there is reason for cautious optimism. For the first time in history, people are beginning to share across borders, because cultures and continents are growing closer together, because physical distance hardly matters anymore, and because knowledge is becoming the most valuable means of production. To pursue only one's own interest becomes more and more perilous and less and less profitable. For by risking their reputations, egocentrics also lessen their chance of success.

Altruists have always profited from being able to exchange information. The more people around the world know about and depend on one another, the higher the benefits and the lower the risks of selflessness.

The Joy of Giving

A WOMAN RINGS YOUR DOORBELL AND ASKS how you are. As thanks for the information, she hands you an envelope containing $50. You may purchase whatever you want with the money, but you must spend it by sundown.

Then the same woman goes to your neighbors' house. They too are asked how they are and receive $50, but they are required to give the money away. They can donate it to a good cause, give it to a beggar, buy a toy for a child, or invite friends to dinner, but again, they must get rid of the money by sundown.

When darkness has fallen, your telephone rings. It is the woman again, asking what you bought and how you feel *now*. She asks your neighbors the same question.

Whom has the money made happier? Does your neighbor have reason to envy you? This good fairy asked many people if it would make them happier to spend $50 on themselves or to give it away. Most answered that they would rather fulfill a wish of their own. Yet when the woman distributed her largesse as described above, no matter what people had answered before, it was always the ones who spent the money on others who were in a better mood that evening.

This is no fairy tale but an experiment conducted by the Canadian psychologist Elizabeth Dunn.[1] The good-fairy researcher asked more than 600 randomly chosen Americans what percentage of their income they spend on gifts and donations and how happy they were. Once again, the more generous responders were happier. Finally, Dunn interviewed employees who had received a bonus of several thousand dollars and either kept it for themselves or given it away. The results were the same.

The experiments show how mistaken most people are about what they need. We assume that of course it will be better for us to have more rather than less money in our pocket. Similarly, we think that more free time will make us happier. But if that were so, Germany, for one, would be an Isle of the Blessed. In fact, the percentage of happy Germans has not risen at all since the postwar period. Numerous studies and even more life histories have disproven the assumption that more money and more free time will make people happier in the long run.[2] On the other hand, many studies confirm the findings of the good fairy Elizabeth Dunn: People who voluntarily do things for others not only feel good at the moment, but also raise their satisfaction with their lives over the long term.[3]

The human brain was programmed at a time when hunger was our forebears' constant companion and giving a gift required a huge sacrifice. Grabbing what you could get often meant the difference between life and death. But today, greed, like the appendix, has lost its evolutionary purpose. Since almost all the inhabitants of the developed countries live in affluence, more possessions do not lead to better reproductive chances. On the contrary, altruists are healthier and live longer, as long as their selflessness does not drive them to the total abandonment of self.

Yet greed, like our unhealthy appetite for fat and sugar, continues to haunt us. The neuropsychologist Jaak Panksepp calls it "a goad without a goal."[4] We want more—more money, more status, more free time, even if they do not make us happier in the end and have no other point of their own. We still slave away to have a two-week vacation in Bermuda, get the corner office, or own a bigger house.

Our society reinforces this programming. From the popular sport of bargain hunting to exorbitant bonuses for executives, the past decades have seen an increasing tendency to motivate people with money. As obvious as the idea seems, it quickly reached the limit of its effectiveness. According to pure economic theory, the invisible hand of the markets ought to transform the egocentrism of individuals into communal good, just as alchemy was supposed to transform lead into gold. Recent crises show the opposite. If individuals are encouraged to be greedy, neither the general welfare nor general well-being is increased.

In reality, no communal human life functions without selflessness, and in the future, altruism will become even more important. We simply can no longer afford to squander this resource. But how can a society encourage people's readiness to care for others? Science has provided us with a flood of new findings about when and why we commit ourselves to others and to common goals. Of course, not every result from the laboratory can be applied to society at large. Nevertheless, the research allows us to draw conclusions about conditions that encourage people's generous predispositions rather than their selfishness. From these insights we can derive the principles of successful communities, be they groups, schools, businesses, or entire societies.

For instance, the dream of the autonomous individual—the widespread idea that everyone holds their own destiny in their hands—seems dubious. The ideal of liberal democracy, to give the individual the greatest possible freedom, is in urgent need of revision. Equally important is to strengthen the bonds of mutual dependence among humans. The time of the lone cowboy is over. The more people need one another and admit their neediness, the sooner they will be prepared to share resources and help one another.

Selfless behavior flourishes whenever there is a strong likelihood that the participants will encounter one another again. The decrease in long-term commitments in the workplace as well as in personal relationships works against that likelihood. If you fail to give people adequate prospects for the future (by hiring them only for a limited time, for instance), you shouldn't be surprised at the price you will pay. Lack of commitment produces a lack of loyalty. A flourishing community favors long-term commitments and interests.

A thriving community promotes cultural diversity, trusts its members, and bets on voluntary cooperation. Coercion and control—but also rewards!—make that volition disappear. Unfairness is equally fatal. Nothing destroys cooperation as fast as the feeling of being exploited. Conversely, people are prepared to forego large personal advantages as long as they are convinced that others are being fair. Only sanctions work against freeloaders. Audible praise for even small contributions to the common good, however, is often more effective than punishing malefactors.

The openness of the public sphere encourages morality because cooperative people profit from their good reputations. Moreover, and more important, altruism is contagious. After all, the large majority of humankind belongs to the species *Homo reciprocans*. They respond in kind to their experience. Even if they are only a witness to someone else helping others, it awakens their own altruism.

That's why the willingness to help can never be in vain. Our relationships are like a resonating body: They amplify everything we do. Benevolence begets new acts of benevolence; interpersonal trust increases.

Many people may not be used to letting themselves be guided by the interests of others. The willingness to do something for others, however, is an attitude that one can practice until it is as natural as riding a bicycle. In time the fear of being exploited fades, and with the courage to give grows the feeling of freedom. The journey begins with curiosity. By experimenting with generosity, we have nothing to lose and much to gain, for selflessness makes us happy and transforms the world.

Notes

INTRODUCTION

1 Quoted in Oliner (2003), 195.

2 Stefan Klein, *The Science of Happiness* (Cambridge: Da Capo Press, 2006).

3 Kessler; Murray & Lopez (1997b); Murray & Lopez (1997a); Seligman (1990); Seligman (1998).

PART I: YOU AND I

CHAPTER 1: THE UNEXPLAINED FRIENDLINESS OF THE WORLD

1 In a letter to the geologist Charles Lyell, 1860.

2 Based on a three-part video interview with Autrey, available at http://www.youtube.com/watch?v=bjjkbTcHnYg.

3 Despite what was said in the media, Autrey never served in the Army. See http://www.navy.mil/search/display.asp?story_id=29707.

4 Boyd and Richerson.

5 Organ donation to anonymous beneficiaries is illegal in Germany.

6 Hitchens.

7 Mother Teresa.

8 Ghiselin, 247.

9 Darwin (1888), 200.

10 Spencer, Herbert, *Social Statistics* (London: John Chapman, 1851), Chapter 28. Now available at: http://oll.libertyfund.org/?option=com_staticxt&staticfile=show.php%3Ftitle=273&chapter=6417&layout=html&Itemid=27.

11 Quoted in Hofstadter, 45.

12 http://www.imdb.com/title/tt0094291/quotes.

13 Dawkins (1989).

14 Kropotkin.

15 Quoted in McElreath and Boyd, 82.

16 Hitchens; Hofstadter; Kropotkin; McLean.

17 Dawkins (1989), 2.

18 Clark, 303.

19 Darwin (1962), 496. Darwin also spoke out publicly against slavery. Might the experience in Brazil have been one of the keys that led to his theory of evolution? There is no doubt that his famous investigation of the beaks of finches on the Galapagos Islands was *not* what moved him to solve the riddle of the origin of species. According to the British historians of science Adrian Desmond and James Moore, the real catalyst for Darwin was this painful memory of the groaning Brazilian slave. Even if their thesis is somewhat too confidently expressed; even if there was not one *single* catalyst for one of the most significant achievements of the scientific method, Desmond and Moore have found solid evidence that Darwin was inspired by the idea of proving the common genealogy of all peoples. For as soon as all men and women can be shown to be siblings, the basis for racism would be destroyed once and for all. (Desmond and Moore).

20 Quoted in Wright, 211.

21 Darwin (1962), 205.

22 Ibid., 228.

23 Darwin (1888), Part I, Chapter 4.

CHAPTER 2: GIVE AND TAKE

1 Khalil Gibran, *Jesus, the Son of Man* (http://gutenberg.net.au/ebooks03/0301451h.html).

2 Rousseau (1995), 196.

3 Approximately a million new titles are published worldwide every year. We assume an average thickness of 0.8 inches and include only books published in the last twenty years.

4 Poundstone, 235.

5 A classic publication in this field is Maynard Smith and Price.

6 Blair.

7 Axelrod (1984); Axelrod (1997); Axelrod & Hamilton.

8 A player will frequently allow an opponent two attempts at cheating, but that is not always the best solution. The number of cooperative moves before the first retribution can be optimized. The best possible strategy depends on the various points awarded for cooperation or confrontation. Also, the number of cooperative moves before the first retribution should be constantly and randomly varied; otherwise, a cooperative strategy can be exploited by a malevolent player, who, for example, would always cheat for two moves and then return to cooperation. Nowak and Sigmund (1992).

9 Under certain circumstances, a somewhat more Machiavellian strategy can allow escape from this cycle. This strategy, called "Pavlov," consists in offering the

opponent cooperation if she herself cooperated on her last move or if both players have chosen conflict. On the other hand, if the opponent has good-naturedly allowed herself to be exploited on the last move, or has herself exploited Pavlov, then Pavlov opts for cooperation. This strategy is only successful, however, if both players make their moves simultaneously without knowing what the other will do. But in real life, both partners are more likely to react in turn to the last move of the other. See Nowak and Sigmund (1995), Wedekind, and Milinski.

10 Ridley.

11 Trivers (1971).

12 Quoted in Hrdy (2009), 133.

CHAPTER 3: BUILDING TRUST

1 For the criticism of Turnbull, see Barth and Turnbull; Knight; and Heine.

2 Turnbull, 284.

3 Harford, 20.

4 Kiyonari et al.

5 Rilling et al.

6 I treat this topic at greater length in *The Science of Happiness* (Klein, 2006).

7 Behrens et al.

8 Singer, Kiebel et al. (2004).

9 Samuelson; Vega-Redondo.

10 Lieberman; Tabibnia and Lieberman.

11 Decety et al. (2004).

12 BBC News Online at http://news.bbc.co.uk/2/hi/1392791.stm.

13 McCabe et al. (1996); McCabe and Smith.

14 King-Casas (2005).

15 King-Casas et al. (2005).

16 Krueger.

17 Bartels and Zeki.

18 Kosfeld et al.

19 https://www.verolabs.com/Default.asp

20 Even the simplest scenarios of game theory—e.g., the repeated prisoner's dilemma—depend on so-called "triggers" that decide whether cooperation will be continued or ended. The Folk theorem, one of the central theorems of game theory, can mathematically prove this proposition in a very general form. For a discussion of the Folk theorem, see Gintis (2009).

21   King-Casas (2008).

22   Goddard.

23   Knack and Keefer.

24   Zak and Knack. Beugelsdijk et al., however, dispute the universal validity of this estimate.

CHAPTER 4: FEELINGS WITHOUT BORDERS

1   Klein (2014).

2   Ibid.

3   Rizzolatti and Craighero; Fogassi et al.

4   Mukamel et al.

5   Ramachandran.

6   Wright, 205.

7   Christakis and Fowler, 2000.

8   Rizzolatti and Craighero.

9   Fadiga et al.; Prather et al.; Welberg.

10   The quotes come from Leonardo's manuscripts TP 68 and BN 2038 20r. I have written at greater length about them in *Leonardo's Legacy* (Klein, 2011). The passage recurs almost word for word in the book on the art of painting by the Renaissance theoretician of aesthetics Leon Battista Alberti: "We painters want to express the effects of the mind through the movement of the limbs." Alberti, 272.

11   Leonardo.

12   I have written more extensively about how emotions are anchored in the body in *The Science of Happiness* (Klein, 2006).

13   The transference from physical gestures functions in a similar way (De Gelder).

14   de Waal (2009).

15   Keysers and Gazzola; Keysers and Perrett.

16   Damasio (1995).

17   Singer, Seymour et al. (2004).

18   Myers.

19   Singer et al. (2006).

20   Darley and Batson, 107. The parable is told in Luke 10:30–37.

21   Lessing et al., letter to Friedrich Nicolai of November 1756.

22   Tankersley et al.

23   Frith and Singer; Mitchell et al.

24 Bischof-Köhler.

25 Warneken and Tomasello (2006).

26 Tomasello (2003); Hare et al.

27 de Waal (2006).

28 Warneken and Tomasello (2007).

29 Klein (2014).

30 Lieberman (2006).

31 They were playing the game known as Battle of the Sexes. Adam and Eve have a joint problem. He suggests his solution and she hers. Both are equally good. Each has to choose which to follow without knowing what the other has chosen. If they choose different solutions, they both lose. Pure logic will obviously get them nowhere. While players in the trust game do better when they can accurately assess their partner, the Battle of the Sexes depends on how well you can predict the moves of the other (Kuo et al.).

CHAPTER 5: THERE IS ONLY ONE LOVE

1 Archer.

2 An American study found such emotional upset in approximately 18 percent of people whose dog had died (Katcher).

3 Companies selling pet food and other pet items in Germany have annual sales in the neighborhood of $4,678,000,000, while veterinarians estimate sales of $2,000,000,000 just for the care of dogs and cats (Ohr and Zeddies). The remarkable figure on Americans' spending comes from market research by the American Pet Products Association. More information can be found at www.americanpetproducts.org/press_industrytrends.asp.

4 However, not all of the genes of the rescuer will survive, since part of the genome of the saved relatives is identical.

5 Neyer.

6 Bowles and Posel.

7 Schroeder.

8 Wilson, E. O. (1975).

9 Segal and Hershberger.

10 Gadagkar (2001); Gadagkar (1997).

11 Davies; Cockburn; Welty.

12 DeBruine (2002).

13 Panksepp.

14 See Donaldson and Young; Panksepp; and Lee. Many details still need to be worked out, however. See e.g. Bancroft; Carter (1992).

15 Uvnäs-Moberg and Eriksson; Lee.

16 Leckman et al.; Panksepp.

17 Young et al.

18 Walum et al.

19 Prichard.

20 Kosfeld et al.

21 Petrovic.

22 Baumgartner.

23 Singer et al. (2008).

24 Domes; Guastella.

25 Israel et al.

26 Damasio (1995).

27 Moll et al.

28 Decety et al. (2009); Moll et al.

29 Harbaugh et al.

30 Hobbes, *Leviathan*, chapter 13.

31 Rousseau (2008); Rousseau (2005).

32 Eisenberger et al.

33 I have written at more length about this topic in Klein (2006). See also the literature cited therein.

34 Zorrilla et al.

35 Allman et al.

36 Berkman; Cacioppo; House; Reblin and Uchino.

37 Rodriguez-Laso; S. L. Brown et al. (2003); W. M. Brown et al.

38 Damasio (1995); Klein (2006).

39 Contrary to what many neo-Darwinists often suggest, e.g. Wright.

PART II: ALL OF US

CHAPTER 6: HUMANS SHARE, ANIMALS DON'T

1 Deacon; Wurz; Cremin, 72.

2 Milo.

3 Fehr, Bernhard, and Rockenbach (2008).

4 Hardin.

5   Trivers (1971); Dawkins (1989); Ridley; Wright.

6   Langford et al.; Dugatkin; Clutton-Brock.

7   Wilkinson.

8   Fehr, "The Economics of Impatience" (2002); Loewenstein et al.

9   Stevens and Hauser.

10 Shermer gives a good overview; see also the literature cited there.

11 Hammerstein; Morell.

12 Packer and Pusey.

13 Stiner et al.

14 Morell; Heinsohn and Packer; Packer and Pusey.

15 Boesch (2005); Mitani and Watts (2001); Gilby; Muller.

16 Ueno and Matsuzawa. This unwillingness of mothers to share food with their children is also present in other great apes, both in captivity and in the wild. See Nowell and Fletcher.

17 Tomasello et al. (2009).

18 Silk et al. (2005); Silk (2006).

19 Inoue and Matsuzawa; Herrmann et al. (2007).

20 Boesch and Boesch-Achermann (2000).

21 Boesch et al. (2010).

22 de Waal (2009).

23 Vrba; Schrenk.

24 McManus; Martrat; Johnsen et al.

25 Zhao et al.

26 Chen and Wen-Hsiung Li.

27 Hrdy.

28 Deacon.

29 Fehr et al., "Egalitarianism" (2008); Brownell et al.

CHAPTER 7: IT'S THE PRINCIPLE OF THE THING

1   Cameron; Güth et al; Nowack et al. (2000) give an overview of the literature on the usual game outcomes.

2   The American Nobel laureate Vernon L. Smith, a pioneer of empirical economics, argues that that's why games like Ultimatum are unnatural. No one accepts as realistic the situation in which the players encounter each other anonymously and then never again. Even if participants knew that that was the situation, they would

ignore the rules of the game and act as if there would be a second or third game. Smith (2007).

3 Bewley; Campbell and Kamlani. For experimental studies, see Fehr and Kirchsteiger (1994) and Fehr and Kirchsteiger (1997).

4 In Heinrich von Kleist's 1811 novella *Michael Kohlhaas*, two horses belonging to the eponymous hero, a sixteenth-century horse trader, are illegally seized by an official. In his unsuccessful attempt to regain them and receive compensation for the expenses he has incurred, Kohlhaas gradually becomes a fanatical and violent opponent of the state. Although he finally gets his horses back, he is executed as a rebel. E. L. Doctorow retold the story—about a black revolutionary named Coalhouse Walker Jr.—in his 1975 novel *Ragtime*.

5 Lind; Tyler and Lind (2005).

6 Henrich (2004).

7 Frank et al. (1993); Frank et al. (1996); Selten and Ockenfels; Frey and Meier (2004a).

8 It is unlikely that a company that pays per delivery attracts more competitive characters, because the companies in the study were located in different cities (personal communication from Ernst Fehr). What bicycle courier would commute from Zurich to Bern to work for a company that pays in a different way? (Burks et al.).

9 Fehr and Gächter (2002).

10 De Quervain et al.

11 Sanfey (2003).

12 Rozin et al.

13 Knoch et al. (2006).

14 Damasio (1995).

15 Gächter et al. (2008).

16 Gürek et al.; Hauert et al.

17 Hauert et al.

18 Boyd et al. (2003).

19 Frey and Jegen (2001).

20 Frey and Meier (2004b).

21 Rees et al.

22 Fehr and Falk (2002).

23 Titmuss.

24 Gneezy and Rustichini (2000). Swiss scientists arrived at similar results: When volunteer work is remunerated, the supply of volunteers dwindles. Frey and Goette (1999).

CHAPTER 8: US AGAINST THEM

1   Hamilton (1964).

2   The Price equation (Price, 1970) describes how an arbitrary property z (e.g. denoting the frequency of a gene related to altruism) varies over generations. Provided that a population can be divided into k groups of n_i individuals each (i=1 . . . k), a z_i can be assigned to each group. Let $w_i = n'_i/n_i$ denote the fitness of group i, where n'_i, n_i are numbers of individuals in successive generations, and let w be the average fitness of the entire population. In the same way, define z and z' as the averages of z_i and z'_i, respectively, taken over groups, and further let $\Delta z = z' - z$. Then the Price equation holds: $w\Delta z = cov(w_i, z_i) + E(w_i \Delta z_i)$, where (w_i, z_i) is the expectation and (w_i, z_i) is the covariance while taking averages over groups. Hence the equation links the increase or decrease of property z within the total population (left-hand side) to how this property is distributed over groups.

The larger the first summand on the right-hand side, the more distinct the groups are: It measures inter-group competition. The second summand is essentially determined by group fitnesses w_i, therefore describing intra-group competitions. Suppose z measures altruism within a group, then for all i: $\Delta z_i < 0$ and $w_i > 0$ holds. Consequently, the terms on the equation's right-hand side will have opposite signs, and altruism thrives if the first summand's absolute is bigger than the second summand's absolute.

3   Frank, Steven (1995).

4   Schwarz, 127.

5   Price and Smith.

6   Schwarz, 130.

7   Ibid., 131.

8   Ibid.

9   Hamilton (1975).

10   Wilson, E. O. (1975).

11   Begley.

12   Even E. O. Wilson, the most prominent sociobiologist who had long rejected group selection, has now changed sides. David Wilson and E. O. Wilson (2008). The by now classic study of newer theories of group selection is D. S. Wilson and Sober (1998).

13   Hamilton (1975); West and Griffin (2007); Lehmann et al. (2007).

14   Assume 100 inhabitants of each village before the disaster. After the famine, twenty-five persons survive in the first village, of whom every fourth is an altruist. Seventy-five persons survive in the second village, and three-fourths of them are altruists. In the total surviving population of 100 persons, (.25 x 25) + (.75 x 75) = 62.5 (rounded up = 63) are altruists.

15   Bowles (2006); Bowles (2009). In the later publication, Bowles assumes that the group competition arises through war between peoples. But this assumption is unnecessary; the model is also valid when the competition comes about through natural

disasters. To trace the development of altruism back to war is problematic because the earliest evidence of violence between peoples comes from the Neolithic, yet the essential character traits of modern humans had almost certainly been formed earlier. A thorough discussion of this question can be found in Hrdy.

16   Krueger et al; Wallace et al.; Cesarini (2009).

17   Horowitz.

18   Fessler (1999); Fessler (2004); Fessler and Haley; Haidt (2003).

19   Rakoczy et al.; Tomasello et al. (2009).

20   Piaget.

21   Another well-documented example of the significance of cultural norms in a conflict between genetically identical ethnic groups is the expansion of the Nuer in the Sudan at the expense of the Dinka (Kelly).

22   Black, 131.

CHAPTER 9: THE EVIL IN GOODNESS

 1   Interallied Commission (1919); Buzanski and the literature quoted therein.

 2   Grandberg and Sarup; Berreby; Trotter.

 3   Sherif et al. (1961).

 4   Ibid.

 5   Sherif et al., 79.

 6   Ibid. 108, 111.

 7   Erev et al.; Bornstein et al.; Gunnthorsdottir and Rapoport (2006); Tan and Bolle.

 8   Klinger and Rebien.

 9   Bernhard et al.

10   Boehm (2000).

11   Takahashi et al. (2009).

12   Brown (1978).

13   William Shakespeare, *Henry VI*, Part 1, Act IV, scene 1.

14   Diamond.

15   Sosis (2000); Sosis and Bressler (2003).

16   Over time, prohibitions can even re-sort the genes of an entire population. For example, if the cultural differences between two neighboring peoples increase, more and more people will reproduce within their own community. After all, who wants to live with a mate whose daily diet they find repulsive? Thus the genotypes develop separately; in a few centuries, certain genes will occur more often among the

members of one group than among those of the other, and that becomes a further factor favoring selflessness within the group. As has been described, altruism is more worthwhile the closer the genetic relationship. Such gene clustering has in fact been measured by anthropologists working with still-existing tribal societies; see Bowles (2009). Even neighboring peoples, for example in Papua New Guinea, differ more in their genetic makeup than Europeans from the most distant corners of the continent differ from one another. Only by closing themselves off from outsiders could each group preserve its genetic characteristics, and only where there was such divergence could genes for altruism survive. Obviously, the benefit of sticking together far outweighed the damage done by destructive rituals. See Boyd and Richerson; Henrich (2009); Richerson and Boyd.

17   Kinzler et al.

18   Ferschtman et al.

19   TV documentary *The Story of Ziad Jarrah*. Canadian Broadcasting Company. October 10, 2001.

20   Atran.

21   Sageman.

22   Judges 16:30.

CHAPTER 10: THE GOLDEN RULE

1   Scheer.

2   How well such policies worked was manifest when the deportation of Jews began. Almost everyone in Germany looked the other way. A comparison with occupied Poland is especially informative. Poland had its own infamous history of anti-Semitism. Most Polish Jews were not nearly as assimilated into the majority population as were German Jews. The German occupiers executed any Pole caught harboring Jews, while Germans who did the same usually were punished with just a few months of so-called "protective custody" in a concentration camp. Yet despite all that, after the war ten times as many Poles as Germans received the title "Righteous among the Nations," which the state of Israel awarded, according to specific criteria, to anyone who ran a personal risk to protect Jews in the Nazi imperium. This indifference among Germans cannot be explained by saying that there were more spies in Germany than elsewhere. Rather, the Nazi regime succeeded in turning Jews into outsiders, while the Germans themselves were outsiders in the countries they occupied. On the risks to Germans who harbored Jews, see Scheer; Kosmala.

3   Oliner (1988); Schroeder. For a similar study that avoids some methodological difficulties of Oliner's pioneering work while coming to the same conclusion, see Midlarsky et al.; Fagin-Jones and Midlarsky.

4   Oliner (2003).

5   Dubs.

6   That is why Peter Bodberg, a pioneer American Orientalist, translated *rén* as "shared humanity."

7   Confucius (1983), 109.

8   Black, 104–105.

9   Leviticus 19:34.

10   Psalm 72:11–12.

11   http://www.yoga-age.com/upanishads/isha.html

12   Babylonian Talmud, tractate Shabbat 31a. [quoted in Steinberg, Avraham, MD. *Encyclopedia of Jewish Medical Ethics*. (Nanuet, NY: Feldheim, 2003).]

13   Cicero (1888), 166.

14   Wilson, David Sloan (2002).

15   Matthew 5:17.

16   Matthew 21:31.

17   Jaspers (1953).

18   Eisenstadt; Roes and Raymond; Norenzayan and Shariff.

19   Henrich et al. (2010).

20   Roes and Raymond.

21   Proverbs 25: 21–22.

22   Second Kings 6:8–23.

23   Second Kings 6:23.

24   Phelps.

25   Lieberman (2005).

26   Confucius (1983), 109.

27   Udana Varga 5, 18. [quoted in Sharp, Michael.]

28   http://www.mahabharataonline.com/translation/mahabharata_13b078.php.

29   Nawawi.

30   Kant, (Book I, Chapter I, § 70).

31   Donner.

32   Nowak and Sigmund (2005).

33   Shakespeare, *Richard II*, act I, scene 1.

34   Takahashi (2000).

35   Milinski et al; Wedekind (2000).

36   Ohtsuki and Iwasa.

37   Ensminger.

CHAPTER 11: THE TRIUMPH OF SELFLESSNESS

1 Stallmann.

2 Imhorst.

3 Tapscott and Williams.

4 The often-heard opinion that Wikipedia authors and developers of free software work mainly to attract attention to themselves and advance their careers is not true. In their very thorough analysis, the American economists Karim R. Lakhani and Robert Wolf were able to show that programmers of free software are intrinsically motivated. Like Stallman, they enjoy being technically creative. See Lakhani and Wolf. Studies of Wikipedia authors came to the same conclusion: Their motivation is mainly the joy of writing and their conviction that they are working for a good cause. See Nov (2007); Nov and Kuk (2008); and the literature cited therein.

5 Bundeszentrale für politische Bildung (German Federal Center for Political Education), 2009.

6 Adam Smith, *Wealth of Nations*, Book I, Chapter 1.

7 Hodgson.

8 Mokyr; Posner.

9 Herrmann et al. (2008).

10 Putnam et al. (1993).

11 That is why the party of the media mogul and politician Silvio Berlusconi could never capture a majority in the Emilia Romagna.

12 For European countries, see Adam; for American states, see Putnam (2001).

13 This number comes from the US Agency for International Development. Full report at http://2001-2009.state.gov/p/sca/rls/fs/2005/58392.htm.

14 Radtke.

15 Pérez.

16 Buchan et al.

17 Bullinger et al.

18 Tapscott and Williams.

19 Millward Brown.

EPILOGUE: THE JOY OF GIVING

1 Dunn et al.

2 See the literature cited in Chapter 15 of Klein (2006).

3 Frey and Stutzer (2007); Post; Meier and Stutzer.

4 Panksepp.

Bibliography

Adam, F. *Social Capital across Europe: Findings, Trends and Methodological Shortcomings of Cross-National Surveys* (Berlin: WZB, 2006).

Alberti, L.B. *Das Standbild. Die Malkunst.* (Darmstadt Wissenschaftliche Buchgesellschaft, 2000).

Allman, John, et al. "Parenting and Survival in Anthropoid Primates: Caretakers Live Longer." *Proceedings of the National Academy of Sciences of the United States of America* 95, no. 12 (1998), 6866–6869.

Archer, John. "Why Do People Love Their Pets?" *Evolution and Human Behavior* 18, no. 4 (1997), 237–259.

Atran, S. "Genesis of Suicide Terrorism." *Science* 299, no. 5612 (3, 2003), 1534–1539.

Axelrod, Robert. *The Complexity of Cooperation: Agent-Based Models of Competition and Collaboration* (Princeton: Princeton University Press, 1997).

_____. *The Evolution of Cooperation* (New York: Basic Books, 1984).

_____ and William D. Hamilton. "The Evolution of Cooperation." *Science* 211 (1981), 1390–1396.

Bancroft, J. "The Endocrinology of Sexual Arousal." *Journal of Endocrinology* 186, no. 3 (2005), 411–427.

Bartels, A. and S. Zeki. "The Neural Correlates of Maternal and Romantic Love." *Neuroimage* 21, no. 3 (2004), 1155–1166.

Barth, F. and C. Turnbull. *On Responsibility and Humanity: Calling a Colleague to Account*, vol. 15.1 (Chicago: University of Chicago Press; Wenner-Gren Foundation for Anthropological Research, 1974).

Baumgartner, T. "Oxytocin Shapes the Neural Circuitry of Trust and Trust Adaptation in Humans." *Neuron* 58, no. 4 (2008), 639–650.

Begley, Sharon. "Why Do We Rape, Kill and Sleep Around?" *Newsweek*, June 20, 2009. http://www.newsweek.com/id/202789/page/1.

Behrens, Timothy E. J. et al. "Associative Learning of Social Value." *Nature* 456, no. 7219 (2008), 245–249.

Berkman, L. F. "Social Networks, Host Resistance, and Mortality: A Nine-Year Follow-Up Study of Alameda County Residents." *American Journal of Epidemiology* 109, no. 2 (1979), 186–204.

Bernhard, Helen, Urs Fischbacher, and Ernst Fehr. "Parochial Altruism in Humans." *Nature* 442, no. 7105 (2006), 912–915.

Berreby, David. *Us and Them.* (Chicago: University of Chicago Press, 2008).

Beugelsdijk, S., H. L. F. De Groot, and A. B. T. M. Van Schaik. "Trust and Economic Growth: A Robustness Analysis." *Oxford Economic Papers* 56, no. 1 (2004), 118.

Bewley, T. F. *Why Wages Don't Fall During a Recession* (Cambridge: Harvard University Press, 2002).

Bischof-Köhler, D. "The Development of Empathy in Infants." In Michael E. Lamb and Heidi Keller, *Infant Development: Perspectives from German-Speaking Countries* (Hillsdale, New Jersey: L. Erlbaum Associates, 1991), 245–273.

Black, Antony. *A World History of Ancient Political Thought* (Oxford: Oxford University Press, 2009).

Blair, Clay, Jr. "Passing of a Great Mind," *Life Magazine*, February 25, 1957.

Boehm, Christopher. *Hierarchy in the Forest: The Evolution of Egalitarian Behavior* (Cambridge: Harvard University Press, 2000).

Boesch, C. and H. Boesch-Achermann. *The Chimpanzees of the Tai Forest: Behavioural Ecology and Evolution.* (Oxford: Oxford University Press, 2000).

Boesch, Christophe. "Joint Cooperative Hunting among Wild Chimpanzees: Taking Natural Observations Seriously." *Behavioral and Brain Sciences* 28, no. 5 (2005), 692–693.

_____, et al. "Altruism in Forest Chimpanzees: The Case of Adoption." ed. by Laurie Santos. *PLoS ONE* 5, no. 1 (2010), e8901.

Bornstein, G., U. Gneezy, and R. Nagel. "The Effect of Intergroup Competition on Group Coordination: An Experimental Study." *Games and Economic Behavior* 41, no. 1 (2002), 1–25.

Bowles, S. and D. Posel. "Genetic Relatedness Predicts South African Migrant Workers' Remittances to Their Families." *Nature* 434, no. 7031 (2005), 380–383.

Bowles, S. "Did Warfare Among Ancestral Hunter-Gatherers Affect the Evolution of Human Social Behaviors?" *Science* (New York) 324, no. 5932 (2009), 1293–1298.

_____. "Group Competition, Reproductive Leveling, and the Evolution of Human Altruism." *Science* 314, no. 5805 (2006), 1569–1572.

Boyd, Robert et al. "The Evolution of Altruistic Punishment." *Proceedings of the National Academy of Sciences of the United States of America* 100, no. 6 (2003), 3531–3535.

Boyd, Robert and Peter J. Richerson. "Culture and the Evolution of Human Cooperation." *Philosophical Transactions of the Royal Society. B: Biological Sciences* 364, no. 1533 (2009), 3281–3288.

Brown, R. "Divided We Fall: An Analysis of Relations between Sections of a Factory Workforce." In Henri Tajfel, *Differentiation between Social Groups: Studies in the Social Psychology of Intergroup Relations* (London and New York: Academic Press, 1978), 395–429.

Brown, Stephanie L. et al. "Providing Social Support May Be More Beneficial Than Receiving It: Results from a Prospective Study of Mortality." *Psychological Science: A Journal of the American Psychological Society* 14, no. 4 (2003), 320–327.

Brown, William Michael, Nathan S. Consedine, and Carol Magai. "Altruism Relates to Health in an Ethnically Diverse Sample of Older Adults." *Journal of Gerontology: Psychological Sciences* 60, no. 3 (2005), 143–152.

Brownell, C. A., M. Svetlova, and S. Nichols. "To Share or Not to Share: When Do Toddlers Respond to Another's Needs?" *Infancy* 14, no. 1 (2009), 117–130.

Buchan, Nancy R. et al. "Globalization and Human Cooperation." *Proceedings of the National Academy of Sciences* 106, no. 11 (2009), 4138–4142.

Buddha. *Reden des Buddha* (Stuttgart: Reclam, 2006).

Bullinger, H.-J., I. Haus, P. Ohlhausen, and C. Wagner. "Produktionsfaktor Wissen" *Personalwirtschaft* 5, no. 98 (1998), 22–26.

Bundeszentrale für politische Bildung. *Globalisierung: Zahlen und Fakten* (Bonn, 2009). http://www.bpb.de/wissen/W0UF4G,0,0,Downloads.html.

Burks, Stephen, Jeffrey Carpenter, and Lorenz Goette. "Performance Pay and Worker Cooperation: Evidence from an Artefactual Field Experiment." *Journal of Economic Behavior & Organization* 70, no. 3 (2009), 458–469.

Buzanski, Peter M. "The Interallied Investigation of the Greek Invasion of Smyrna, 1919." *Historian* 25, no. 3 (1963), 325–343.

Cacioppo, John. "Loneliness as a Specific Risk Factor for Depressive Symptoms: Cross-Sectional and Longitudinal Analyses." *Psychology and Aging* 21, no. 1 (2006), 140.

Cameron, L. A. "Raising the Stakes in the Ultimatum Game: Experimental Evidence from Indonesia." *Economic Inquiry* 37, no. 1 (1999), 47–59.

Campbell, C. M. III and K. S. Kamlani. "The Reasons for Wage Rigidity: Evidence from a Survey of Firms." *Quarterly Journal of Economics* 112, no. 3 (1997), 759–789.

Carter, C. S. "Oxytocin and Sexual Behavior." *Neuroscience and Biobehavioral Reviews* 16, no. 2 (1992), 131–144.

Cesarini, David. "Genetic Variation in Preferences for Giving and Risk Taking." *Quarterly Journal of Economics* 124, no. 2 (2009), 809–842.

Chen, Feng-Chi and Wen-Hsiung Li. "Genomic Divergences between Humans and Other Hominoids and the Effective Population Size of the Common Ancestor of Humans and Chimpanzees." *American Journal of Human Genetics* 68, no. 2 (2001), 444–456.

Christakis, Nicholas and James H. Fowler. *Connected: The Amazing Power of Social Networks and How They Shape Our Lives* (London: Harperpress, 2000).

Cicero, Marcus Tullius. *Tusculan Disputations*. Trans. by C. D. Yonge. (New York, Harper & Bros., 1888), http://archive.org/stream/cicerostusculand00ciceuoft#page/n5/mode/2up.

_____. *Über die Gesetze / Stoische Paradoxien* (Düsseldorf: Artemis & Winkler, 1994).

Ronald Clark, J. B. S.: The Life and Work of J. B. S. Haldane (New York: Oxford University Press, 1984), 303.

Clutton-Brock, Tim. "Cooperation between Non-Kin in Animal Societies." *Nature* 462, no. 7269 (2009), 51–57.

Cockburn, A. "Why Do So Many Australian Birds Cooperate: Social Evolution in the Corvida?" In: Floyd, R. B., A. W. Sheppard, P. J. De Barro, eds. *Frontiers of Population Ecology.* (Melbourne: CSIRO Publishing, 1996), 451–472.

Confucius (Konfuzius). *Gespräche = Lun-yu*, 3rd ed. (Munich: Beck, 2007).

_____. *The Analects* (Hong Kong: Chinese University Press, 1983).

Cremin, Aedeen, ed. *The World Encyclopedia of Archeology* (Buffalo: Firefly Books, 2007).

Damasio, Antonio. *Descartes' Irrtum: Fühlen, Denken und das menschliche Gehirn* (Munich and Leipzig: List, 1995).

Darley, John M. and C. Daniel Batson. "'From Jerusalem to Jericho': A Study of Situational and Dispositional Variables in Helping Behavior." *Journal of Personality and Social Psychology*, Vol. 27 (1973), 100–108.

Darwin, Charles. *The Descent of Man and Selection in Relation to Sex*, 2nd ed. (London: John Murray, 1888).

_____. *The Voyage of the Beagle* (Garden City: Doubleday / Anchor Books, 1962).

_____. *Origin of Species*, online variorum edition: http://darwin-online.org.uk/Variorum/1859/1859-1-dns.html.

Davies, N. "Nestling Cuckoos, *Cuculus canorus*, Exploit Hosts with Begging Calls That Mimic a Brood." *Proceedings of the Royal Society B: Biological Sciences* 265, no. 1397 (1998), 673.

Dawkins, Richard. *The Selfish Gene*, 2nd ed. (Oxford: Oxford University Press, 1989).

Deacon, H. "The Stratigraphy and Sedimentology of the Main Site Sequence, Klasies River, South Africa." *The South African Archaeological Bulletin* 43, No. 147 (1988), 5.

DeBruine, Lisa M. "Facial Resemblance Enhances Trust." *Proceedings of the Royal Society B: Biological Sciences* 269, no. 1498 (2002), 1307–1312.

Decety, J. et al. "The Neural Bases of Cooperation and Competition: An fMRI Investigation." *Neuroimage* 23, no. 2 (2004), 744–751.

_____. *The Social Neuroscience of Empathy* (Cambridge: MIT Press, 2009).

De Gelder, B. "Towards the Neurobiology of Emotional Body Language." *Nature Reviews Neuroscience* 7, no. 3 (2006), 242–249.

De Quervain, D. J. F. et al. "The Neural Basis of Altruistic Punishment." *Science* 305, no. 5688 (2004), 1254.

Desmond, Adrian and James Moore. *Darwin's Sacred Cause: How a Hatred of Slavery Shaped Darwin's Views on Human Evolution* (Boston: Houghton Mifflin Harcourt, 2009).

Deutschkron, Inge. *Ich trug den gelben Stern* (Munich: Deutscher Taschenbuch Verlag, 1992).

Diamond, Jared. *Collapse: How Societies Choose to Fail or Succeed*, rev. ed. (New York: Penguin, 2011).

Domes, Gregor. "Oxytocin Improves 'Mind-Reading' in Humans." *Biological Psychiatry* 61, no. 6 (2007), 731.

Donaldson, Zoe R. and Larry J. Young. "Oxytocin, Vasopressin, and the Neurogenetics of Sociality." *Science* 322, no. 5903 (2008), 900–904.

Donner, Herbert. *Geschichte des Volkes Israel und seiner Nachbarn in Grundzügen* (Göttingen: Vandenhoeck & Ruprecht, 1986).

Dubs, H. H. "The Development of Altruism in Confucianism." *Philosophy East and West* 1, no. 1 (1951), 48–55.

Dugatkin, L. A. *Cooperation among Animals: An Evolutionary Perspective* (Oxford: Oxford University Press, 1997).

Dunn, E. W., L. B. Aknin, and M. I. Norton. "Spending Money on Others Promotes Happiness." *Science* 319, no. 5870 (2008), 1687–1688.

Eisenberger, Naomi I., Matthew D. Lieberman, and Kipling D. Williams. "Does Rejection Hurt? An FMRI Study of Social Exclusion." *Science* 302, no. 5643 (2003), 290–292.

Eisenstadt, Shmuel. *Kulturen der Achsenzeit: ihre Ursprünge und ihre Vielfalt* (Frankurt am Main: Suhrkamp, 1987).

Ensminger, J. "Transaction Costs and Islam: Explaining Conversion in Africa." *Journal of Institutional and Theoretical Economics* 153, no. 1 (1997).

Erev, I., G. Bornstein, and R. Galili. "Constructive Intergroup Competition as a Solution to the Free Rider Problem: A Field Experiment." *Journal of Experimental Social Psychology* 29, no. 6 (1993), 463–478.

Fadiga, Luciano, Laila Craighero, and Alessandro D'Ausilio. "Broca's Area in Language, Action, and Music." *Annals of the New York Academy of Sciences* 1169 (2009), 448–458.

Fagin-Jones, S. and E. Midlarsky. "Courageous Altruism: Personal and Situational Correlates of Rescue during the Holocaust." *The Journal of Positive Psychology* 2, no. 2 (2007), 136–147.

Fehr, E. and Simon Gächter. "Altruistic Punishment in Humans." *Nature* 415, no. 6868 (2002), 137.

Fehr, E., Helen Bernhard, and Bettina Rockenbach. "Egalitarianism in Young Children." *Nature* 454, no. 7208 (2008), 1079–1083.

Fehr, E. and G. Kirchsteiger. "Insider Power, Wage Discrimination and Fairness." *The Economic Journal* (1994), 571–583.

Fehr, E. and A. Falk. "Psychological Foundations of Incentives." *European Economic Review* 46, no. 4 (2002), 687–724.

Fehr, E. and G. Kirchsteiger. "Reciprocity as a Contract Enforcement Device: Experimental Evidence." *Econometrica* 65, no. 4 (1997), 833–860.

Fehr, E. "The Economics of Impatience." *Nature* 415, no. 6869 (2002), 268–269.

Ferschtman, C., U. Gneezy, and F. Verboven. "Discrimination and Nepotism: The Efficiency of the Anonymity Rule." *The Journal of Legal Studies* 34, no. 2 (2005), 371–396.

Fessler, D. M. T. "Toward an Understanding of the Universality of Second Order Emotions." *Biocultural Approaches to the Emotions* (1999), 75.

_____. "Shame in Two Cultures: Implications for Evolutionary Approaches." *Journal of Cognition and Culture* 4, no. 2 (2004), 207–262.

_____ and K. J. Haley. "The Strategy of Affect: Emotions in Human Cooperation." In Peter Hammerstein, *The Genetic and Cultural Evolution of Cooperation* (Cambridge: MIT Press and Dahlem University Press, 2003), 7–36.

Fogassi, Leonardo et al. "Parietal Lobe: From Action Organization to Intention Understanding." *Science* 308, no. 5722 (2005), 662–667.

Frank, R. H., T. Gilovich, and D. T. Regan. "Do Economists Make Bad Citizens?" *The Journal of Economic Perspectives* (1996), 187–192.

_____. "Does Studying Economics Inhibit Cooperation?" *The Journal of Economic Perspectives* 7, no. 2 (1993), 159–171.

Frank, Steven. "George Price's Contributions to Evolutionary Genetics." *Journal of Theoretical Biology* 175, no. 3 (1995), 373–388.

Frey, B. S. and L. Goette. "Does Pay Motivate Volunteers?" Unpublished manuscript (University of Zurich: Institute for Empirical Economic Research, 1999).

Frey, B. S. and R. Jegen. "Motivation Crowding Theory." *Journal of Economic Surveys* 15, no. 5 (2001), 589–611.

Frey, B. S. and S. Meier. "Pro-Social Behavior in a Natural Setting." *Journal of Economic Behavior and Organization* 54, no. 1 (2004), 65–88.

_____. "Social Comparisons and Pro-Social Behavior: Testing 'Conditional Cooperation' in a Field Experiment." *American Economic Review* (2004), 1717–1722.

Frey, Bruno S. and Alois Stutzer. *Economics and Psychology: A Promising New Cross-Disciplinary Field* (Cambridge: The MIT Press, 2007).

Frith, Chris and Tania Singer. "The Role of Social Cognition in Decision Making." *Philosophical Transactions of the Royal Society B: Biological Sciences* 363, no. 1511 (2008), 3875–3886.

Gächter, Simon et al. "Who Makes a Good Leader? Social Preferences and Leading-by-Example." *CeDEx Discussion Paper* (University of Nottingham), 2008–2016 (2008), http://www.nottingham.ac.uk/economics/cedex/papers/index.html.

Gadagkar, Raghavendra. *Survival Strategies: Cooperation and Conflict in Animal Societies* (Cambridge: Harvard University Press, 1997).

_____. *The Social Biology of "Ropalidia Marginata": Toward an Understanding of the Evolution of Eusociality* (Cambridge: Harvard University Press, 2001).

Ghiselin, Michael T. *The Economy of Nature and the Evolution of Sex* (Berkeley: University of California Press, 1974).

Gibran, Kahlil. *Jesus the Son of Man* (New York: Knopf, 1995).

Gilby, Ian. "Meat Sharing among the Gombe Chimpanzees: Harassment and Reciprocal Exchange." *Animal Behaviour* 71, no. 4 (2006), 953.

Gintis, Herbert. *The Bounds of Reason* (Princeton: Princeton University Press, 2009).

Gneezy U. and A. Rustichini. "Pay Enough or Don't Pay at All." *Quarterly Journal of Economics* 115, no. 3 (2000), 791–810.

Goddard, Roger. "Relational Networks, Social Trust, and Norms: A Social Capital Perspective on Students' Chances of Academic Success." *Educational Evaluation and Policy Analysis* 25, no. 1 (2003), 59.

Gould, Stephen Jay. "Kropotkin Was No Crackpot." *Natural History* 106 (1997), 12–21.

Granberg, Donald and Gian Sarup. *Social Judgment and Intergroup Relations* (New York: Springer, 1992).

Guastella, Adam. "Oxytocin Increases Gaze to the Eye Region of Human Faces." *Biological Psychiatry* 63, no. 1 (2008), 3.

Gunnthorsdottir, A. and A. Rapoport. "Embedding Social Dilemmas in Intergroup Competition Reduces Free-Riding." *Organizational Behavior and Human Decision Processes* 101, no. 2 (2006), 184–199.

Gürek, O., B. Irlenbusch, and B. Rockenbach. "The Competitive Advantage of Sanctioning Institutions." *Science* 312, no. 5770 (2006), 108.

Güth, Werner, Rolf Schmittberger, and Bernd Schwarze. "An Experimental Analysis of Ultimatum Bargaining." *Journal of Economic Behavior & Organization* 3, no. 4 (1982), 367–388.

Haidt, J. "The Moral Emotions." In Richard J. Davidson and Klaus R. Scherer, *Handbook of Affective Sciences* (Oxford: Oxford University Press, 2003), 852–870.

Hamilton, William D. "Innate Social Aptitudes of Man: An Approach from Evolutionary Genetics." *Biosocial Anthropology* (1975), 133–155.

_____. "The Evolution of Altruistic Behavior," *American Naturalist* 97 (1963), 354–356.

_____. "The Genetical Evolution of Social Behaviour." *Journal of Theoretical Biology* 7, no. 1 (1964), 1–52.

Hammerstein, Peter, ed. *Genetic and Cultural Evolution of Cooperation* (Cambridge; MIT Press, 2003).

Harbaugh, W. T., U. Mayr, and D. R. Burghart. "Neural Responses to Taxation and Voluntary Giving Reveal Motives for Charitable Donations." *Science* 316, no. 5831 (2007), 1622–1625.

Hardin, Garrett. "The Tragedy of the Commons." *Science* (1968), 1243–1248.

Hare, B. et al. "Do Capuchin Monkeys, *Cebus apella*, Know What Conspecifics Do and Do Not See?" *Animal Behaviour* 65, no. 1 (2003), 131–142.

Harford, Tim. *Die Logik des Lebens: Warum Ihr Boss überbezahlt ist, Oralsex boomt und New Orleans nicht wieder aufgebaut wird—Die rationalen Motive unserer scheinbar irrationalen Entscheidungen* (Munich: Riemann Verlag, 2008).

Hauert, C. et al. "Via Freedom to Coercion: The Emergence of Costly Punishment." *Science* 316, no. 5833 (2007), 1905–1907.

Heine, B. "The Mountain People: Some Notes on the Ik of North-Eastern Uganda." *Africa: Journal of the International African Institute* 55, no. 1 (1985), 3–16.

Heinsohn, R. and Craig Packer. "Complex Cooperative Strategies in Group-Territorial African Lions." *Science* 269, no. 5228 (1995), 1260.

Henrich, J. "The Evolution of Costly Displays, Cooperation and Religion: Credibility Enhancing Displays and Their Implications for Cultural Evolution." *Evolution and Human Behavior* 30, no. 4 (2009), 244–260.

Henrich, J. et al. "Markets, Religion, Community Size, and the Evolution of Fairness and Punishment." *Science* 327, no. 5972 (2010), 1480–1484.

Henrich, Joseph Patrick. *Foundations of Human Sociality* (Oxford: Oxford University Press, 2004).

Herrmann, B., C. Thöni, and S. Gachter. "Antisocial Punishment across Societies." *Science* 319, no. 5868 (2008), 1362–1367.

Herrmann, Esther et al. "Humans Have Evolved Specialized Skills of Social Cognition: The Cultural Intelligence Hypothesis." *Science* 317, no. 5843 (2007), 1360–1366.

Hitchens, Christopher. *The Missionary Position: Mother Teresa in Theory and Practice* (London: Verso, 1995)

Hobbes, Thomas. *Leviathan* (London: Andrew Crooke, 1651), facsimile at http://archive.org/stream/hobbessleviathan00hobbuoft#page/n35/mode/2up.

Hodgson, Bernard, ed. *The Invisible Hand and the Common Good.* (Berlin, Heidelberg: Springer-Verlag, 2004).

Hofstadter, Richard. *Social Darwinism in American Thought*, revised edition (New York: George Braziller, 1959).

Horowitz, A. "Disambiguating the Guilty Look." *Behavioural Processes* 81, no. 3 (2009), 447–452.

House, J. S. "Social Relationships and Health." *Science* 241, no. 4865 (1988), 540–545.

Hrdy, Sarah Blaffer. *Mothers and Others: The Evolutionary Origins of Mutual Under-standing* (Cambridge: The Belknap Press of Harvard University Press, 2009).

Imhorst, C. *Die Anarchie der Hacker* (Marburg: Tectum-Verlag, 2004).

Inoue, S. and T. Matsuzawa. "Working Memory of Numerals in Chimpanzees." *Current Biology* 17, no. 23 (12, 2007), R1004–R1005.

Interallied Commission of Enquiry into the Greek Occupation of Smyrna and Adjacent Territories. *Rapport de la Commission interalliée d'enquête sur l'occupation grecque de Smyrne et des territoires adjacents* (Paris, 1919).

Isocrates. *Sämtliche Werke* (Stuttgart: A. Hiersemann, 1993).

Israel, Solomon et al. "The Oxytocin Receptor (OXTR) Contributes to Prosocial Fund Allocations in the Dictator Game and the Social Value Orientations Task." *PLoS ONE* 4, no. 5 (2009), e5535.

Jaspers, Karl. *The Origin and Goal of History*. Trans. by Michael Bullock. (New Haven: Yale University Press, 1953).

Johnsen, S. J., W. Dansgaard, and J. W. C. White. "The Origin of Arctic Precipitation under Present and Glacial Conditions." *Tellus. Series B, Chemical and Physical Meteorology* 41, no. 4 (1989), 452–468.

Kant, Immanuel. *Critique of Practical Reason*. Trans. by Thomas Kingsmill Abbott (Project Gutenberg, 2004), http://www.gutenberg.org/cache/epub/5683/pg5683.txt.

Katcher, Aaron and International Conference on the Human-Companion Animal Bond. *New Perspectives on Our Lives with Companion Animals* (Philadelphia: University of Pennsylvania Press, 1983).

Kelly, Raymond. *The Nuer Conquest* (Ann Arbor: University of Michigan Press, 1985). http://openlibrary.org/b/OL3021244M/Nuer_conquest.

Kessler, Ronald C. "The Epidemiology of Major Depressive Disorder: Results from the National Comorbidity Survey Replication (NCS-R)." *Journal of the American Medical Association* 289, no. 23 (2003), 3095–3105.

Keysers, C. and V. Gazzola. "Towards a Unifying Neural Theory of Social Cognition." *Understanding Emotions* (2006), 379.

Keysers C. and D. I. Perrett. "Demystifying Social Cognition: A Hebbian Perspective." *Trends in Cognitive Sciences* 8, no. 11 (2004), 501–507.

King-Casas, Brooks et al. "Getting to Know You: Reputation and Trust in a Two-Person Economic Exchange." *Science* 308, no. 5718 (April 1, 2005), 78–83.

King-Casas, B. "The Rupture and Repair of Cooperation in Borderline Personality Disorder." *Science* 321, no. 5890 (8, 2008), 806–810.

Kinzler, Katherine D., Emmanuel Dupoux, and Elizabeth S. Spelke. "The Native Language of Social Cognition." *Proceedings of the National Academy of Sciences* 104, no. 30 (2007), 12577–12580.

Kiyonari, T., S. Tanida, and T. Yamagishi. "Social Exchange and Reciprocity: Confusion or a Heuristic?" *Evolution and Human Behavior* 21, no. 6 (2000), 411–427.

Klein, Stefan. *Leonardo's Legacy* (Cambridge: Da Capo Press, 2011).

_____. *The Science of Happiness* (Cambridge: Da Capo Press, 2006).

_____. *We Are All Stardust*. Translated by Ross Benjamin. (New York: The Experiment, Forthcoming 2014).

Klinger, Sabine and Martina Rebien. *Soziale Netzwerke helfen bei der Personalsuche*. IAB Kurzbericht (Nürnberg: Institut für Arbeitsmarkt- und Berufsforschung, 2009).

Knack, Stephen and Philip Keefer. "Does Social Capital Have an Economic Payoff? A Cross-Country Investigation." *Quarterly Journal of Economics* 112, no. 4 (1997), 1251–1288.

Knight, J. "'The Mountain People' as Tribal Mirror." *Anthropology Today* (1994), 1–3.

Knoch, D. et al. "Diminishing Reciprocal Fairness by Disrupting the Right Prefrontal Cortex." *Science* 314, no. 5800 (2006), 829.

Kosfeld, Michael, Markus Heinrichs, Paul J. Zak, Urs Fischbacher, and Ernst Fehr. "Oxytocin Increases Trust in Humans." *Nature* 435, no. 7042 (2005), 673–676.

Kosmala, Beate. *Solidarität und Hilfe für Juden während der NS-Zeit. Hilfe für Juden in Deutschland, 1941–1945* (Berlin: Metropol, 2002).

Kropotkin, Peter. *Gegenseitige Hilfe* (Frafenau: Trotzdem Verlag, 2005).

Krueger, Frank et al. "Neural Correlates of Trust." *Proceedings of the National Academy of Sciences* 104, no. 50 (2007), 20084–20089.

Kuo, W. J., T. Sjostrom, Y. P. Chen, Y. H. Wang, and C. Y. Huang. "Intuition and Deliberation: Two Systems for Strategizing in the Brain." *Science* 324, no. 5926 (2009), 519.

Lakhani, K. and R. G. Wolf. "Why Hackers Do What They Do: Understanding Motivation and Effort in Free / Open Source Software Projects." In Joseph Feller et al., *Perspectives on Free and Open Source Software* (Cambridge: MIT Press, 2005).

Lang, F. "Social Relationships as Nature and Nurture: An Evolutionary-Psychological Framework of Social Relationship Regulation." *Zeitschrift für Soziologie der Erziehung und Sozialisation* 25 (2005), 162–177.

Langford, D. J., S. E. Crager, Z. Shehzad, S. B. Smith, S. G. Sotocinal, J. S. Levenstadt, M. L. Chanda, D. J. Levitin, and J. S. Mogil. "Social Modulation of Pain as Evidence for Empathy in Mice." *Science* 312, no. 5782 (2006), 1967.

Leckman, James F., Wayne K. Goodman, William G. North, Phillip B. Chappell, Lawrence H. Price, David L. Pauls, George M. Anderson, et al. "The Role of Central Oxytocin in Obsessive Compulsive Disorder and Related Normal Behavior." *Psychoneuroendocrinology* 19, no. 8 (1994), 723–749.

Lee, Heon-Jin. "Oxytocin: The Great Facilitator of Life." *Progress in Neurobiology* 88, no. 2 (2009), 127.

Lehmann, Laurent, Laurent Keller, and Stuart West. "Group Selection and Kin Selection: Two Concepts but One Process." *Proceedings of the National Academy of Sciences of the United States of America* 104, no. 16 (2007), 6736.

Leimar, O. and P. Hammerstein. "Evolution of Cooperation through Indirect Reciprocity." *Proceedings of the Royal Society of London, Series B: Biological Sciences* 268, no. 1468 (2001), 745–753.

Leonardo Da Vinci. *Das Buch von der Malerei*, German edition after the *Codex Vaticanus, 1270* (Stuttgart: Kohlhammer, 1885).

Lessing G. E., M. Mendelssohn, and F. Ficolai. *Briefwechsel über das Trauerspiel*, ed. by J. Schulte-Sasse (Munich: Winkler Verlag, 1972).

Lieberman, M. D. "Social Cognitive Neuroscience: A Review of Core Processes." *Annual Review of Psychology* 58 (2006), 259–289.

_____ and A. Hariri, J. M. Jarcho, and S. Y. Bookheimer. "An fMRI Investigation of Race-Related Amygdala Activity in African-American and Caucasian-American Individuals." *Nature Neuroscience* 8, no. 6 (2005), 720–722.

Lind, E. "Individual and Corporate Dispute Resolution: Using Procedural Fairness as a Decision Heuristic." *Administrative Science Quarterly* 38, no. 2 (1993), 224.

Loewenstein, George, Daniel Read, and Roy F. Baumeister. *Time and Decision* (Russell Sage Foundation, 2003).

Martrat, Belen. "Abrupt Temperature Changes in the Western Mediterranean over the Past 250,000 Years." *Science* 306, no. 5702 (2004), 1762.

Maynard Smith, J. and G. R. Price. "The Logic of Animal Conflict." *Nature* 246, no. 5427 (1973), 15–18.

McCabe, Kevin A. and Vernon L. Smith. "A Comparison of Naïve and Sophisticated Subject Behavior with Game Theoretic Predictions." *Proceedings of the National Academy of Sciences of the United States of America* 97, no. 7 (2000), 3777–3781.

_____ and Stephen J. Rassenti. "Game Theory and Reciprocity in Some Extensive Form Experiment Games." *Proceedings of the National Academy of Sciences of the United States of America* 93, no. 23 (1996), 13421–13428.

McElreath, Richard and Robert Boyd. *Mathematical Models of Social Evolution: A Guide for the Perplexed* (Chicago: University of Chicago Press, 2007).

McLean, Bethany. *The Smartest Guys in the Room: The Amazing Rise and Scandalous Fall of Enron* (New York: Portfolio, 2003).

McManus, J. "A 0.5-Million-Year Record of Millennial-Scale Climate Variability in the North Atlantic." *Science* 283, no. 5404 (1999), 971.

Meier, Stephan and Alois Stutzer. "Is Volunteering Rewarding in Itself?" *Economica* 75, no. 297 (2008), 39–59.

Midlarsky, E., S. Fagin Jones, and R. Nemeroff. "Heroic Rescue during the Holocaust: Empirical and Methodological Perspectives." In *Measurement, Methodology, and Evaluation* (Washington D.C.: American Psychological Association, 2006), 29–45.

Milinski, Manfred, Dirk Semmann, and Hans-Jürgen Krambeck. "Donors to Charity Gain in Both Indirect Reciprocity and Political Reputation." *Proceedings of the Royal Society B: Biological Sciences* 269, no. 1494 (5, 2002), 881–883.

Millward Brown. *Top 100: Most Valuable Global Brands 2010* (New York: Millward Brown, 2010). http://www.millwardbrown.com/Libraries/Optimor_BrandZ_Files/2010_BrandZ_Top100_Report.sflb.ashx.

Milo, Richard G. "Evidence for Hominid Predation at Klasies River Mouth, South Africa, and Its Implications for the Behaviour of Early Modern Humans." *Journal of Archaeological Science* 25, no. 2 (1998), 99–133.

Mitani, John C. and David P. Watts. "Why Do Chimpanzees Hunt and Share Meat?" *Animal Behaviour* 61, no. 5 (2001), 915–924.

Mitchell, J. P., M. R. Banaji, and C. N. MacRae. "The Link between Social Cognition and Self-Referential Thought in the Medial Prefrontal Cortex." *Journal of Cognitive Neuroscience* 17, no. 8 (2005), 1306–1315.

Mokyr, Joel. *The Enlightened Economy: An Economic History of Britain 1700–1850* (New Haven: Yale University Press, 2009).

Moll, Jorge et al. "Human Fronto-Mesolimbic Networks Guide Decisions about Charitable Donation." *Proceedings of the National Academy of Sciences of the United States of America* 103, no. 42 (2006), 15623–15628.

Morell, V. "Cowardly Lions Confound Cooperation Theory." *Science* 269, no. 5228 (September 2, 1995), 1216–1217.

Mother Teresa. *Come Be My Light: The Private Writings of the "Saint of Calcutta"*, ed. Brian Kolodiejchuk (New York: Doubleday, 2007).

Mukamel, Roy et al. "Single-Neuron Responses in Humans during Execution and Observation of Actions." *Current Biology* 20 (2010), 1–7.

Muller, M. "Conflict and Cooperation in Wild Chimpanzees." *Advances in the Study of Behavior* 35 (2005), 275–332.

Murray, C. J. L. and A. D. Lopez. "Alternative Projections of Mortality and Disability by Cause 1990–2020: Global Burden of Disease Study." *The Lancet* 349, no. 9064 (1997), 1498–1504.

_____. "Global Mortality, Disability, and the Contribution of Risk Factors: Global Burden of Disease Study." *The Lancet* 349, no. 9063 (1997), 1436–1442.

Myers, David G. *Social Psychology*, 9th ed. (New York: McGraw-Hill Higher Education, 2006).

Nawawi, al-. *Das Buch der Vierzig Hadithe: Kitab al-Arba'in*, ed. Marco Schöller (Frankfurt am Main: Verlag der Weltreigionen, 2007).

Neyer, F. J. "Blood Is Thicker than Water: Kinship Orientation across Adulthood." *Journal of Personality and Social Psychology* 84, no. 2 (2003), 310–321.

Norenzayan, A. and A. F. Shariff. "The Origin and Evolution of Religious Prosociality." *Science* 322, no. 5898 (2008), 58.

Nov, Oded. "What Motivates Wikipedians?" *Communications of the ACM* 50, no. 11 (2007), 64.

_____ and George Kuk. "Open Source Content Contributors' Response to Free-Riding: The Effect of Personality and Context." *Computers in Human Behavior* 24, no. 6 (2008), 2848–2861.

Nowak, M. A. and K. Sigmund. "Invasion Dynamics of the Finitely Repeated Prisoner's Dilemma." *Games and Economic Behavior* 11, no. 2 (1995), 364–390.

_____. "Tit for Tat in Heterogeneous Populations." *Nature* 355, no. 6357 (1992), 250–253.

Nowak, Martin A. and Karl Sigmund. "Evolution of Indirect Reciprocity." *Nature* 437, no. 7063 (10, 2005), 1291–1298.

Nowell, Angela and Alison Fletcher. "Food Transfers in Immature Wild Western Lowland Gorillas (*Gorilla gorilla gorilla*)." *Primates* 47, no. 4 (2006), 294–299.

Ohr, R. and G. Zeddies. "Ökonomische Gesamtbetrachtung der Hundehaltung in Deutschland." *Studie der Wirtschaftswissenschaften* (Göttingen: Universität Göttingen, 2006), 1–35.

Ohtsuki, Hisashi and Yoh Iwasa. "The Leading Eight: Social Norms That Can Maintain Cooperation by Indirect Reciprocity." *Journal of Theoretical Biology* 239, no. 4 (2006), 435–444.

Oliner, Samuel P. *Do Unto Others: Extraordinary Acts of Ordinary People* (Boulder, Colorado: Westview Press, 2003).

_____. *The Altruistic Personality: Rescuers of Jews in Nazi Europe* (New York: Free Press, 1988).

Packer, Craig and Anne E. Pusey. "Divided We Fall: Cooperation among Lions." *Scientific American* 276, no. 5 (1997), 52–59.

Panksepp, Jaak. *Affective Neuroscience: The Foundations of Human and Animal Emotions* (New York: Oxford University Press, 1998).

Parr, L. A. "Understanding Others' Emotions: From Affective Resonance to Empathic Action." *Behavioral and Brain Sciences* 25, no. 1 (2003), 44–45.

Pérez, Maria Camilla. "Facebook Brings Protest to Colombia." *New York Times*, February 8, 2008.

Petrovic, P. "Oxytocin Attenuates Affective Evaluations of Conditioned Faces and Amygdala Activity." *Journal of Neuroscience* 28, no. 26 (June 25, 2008), 6607–6615.

Phelps, E. A. et al. "Performance on Indirect Measures of Race Evaluation Predicts Amygdala Activation." *Journal of Cognitive Neuroscience* 12, no. 5 (2000), 729–738.

Piaget, Jean. *The Moral Judgment of the Child* (Red Lion, PA: Ulan Press, 2012).

Posner, Eric A. *Law and Social Norms* (Cambridge: Harvard University Press, 2002).

Post, Stephen Garrard. *Altruism and Health: Perspectives from Empirical Research* (Oxford: Oxford University Press, 2007).

Poundstone, William. *Prisoner's Dilemma* (New York: Doubleday, 1992).

Prather, J. F. et al. "Precise Auditory-Vocal Mirroring in Neurons for Learned Vocal Communication." *Nature* 451, no. 7176 (2008), 305–310.

Price, George. "Selection and Covariance." *Nature* 227 (1970), 520–521.

Price, G. R. and J. Maynard Smith. "The Logic of Animal Conflict." *Nature* 246 (1973), 15–18.

Prichard, Z. "AVPR1A and OXTR Polymorphisms Are Associated with Sexual and Reproductive Behavioral Phenotypes in Humans." *Human Mutation* 28, no. 11 (2007), 1150.

Putnam, Robert D. *Bowling Alone: The Collapse and Revival of American Community* (New York: Touchstone Books / Simon and Schuster, 2001).

_____, R. Leonardi, and R. Y. Nanetti. *Making Democracy Work: Civic Traditions in Modern Italy* (Princeton: Princeton University Press, 1993).

Radtke, Katrin. *Ein Trend zu transnationaler Solidarität. Die Entwicklung des Spendenaufkommens für die Katastrophen—und Entwicklungshilfe in Deutschland*. WZB Discussion Paper (Berlin: Wissenschaftszentrum Berlin für Sozialforschung, 2007).

Rakoczy, H., F. Warneken, and M. Tomasello. "The Sources of Normativity: Young Children's Awareness of the Normative Structure of Games." *Developmental Psychology* 44, no. 3 (2008), 875–881.

Ramachandran, V. S. "Mirror Neurons and Imitation Learning as the Driving Force behind 'the Great Leap Forward' in Human Evolution." Edge.org, 2006 (http://www.edge.org/3rd_culture/ramachandran/ramachandran_p1.html).

Rand, D. G. et al. "Positive Interactions Promote Public Cooperation." *Science* 325, no. 5945 (2009), 1272–1275.

Reblin, Maija and Bert N. Uchino. "Social and Emotional Support and Its Implication for Health." *Current Opinion in Psychiatry* 21, no. 2 (2008), 201–205.

Rees, M. A. et al. "A Nonsimultaneous, Extended, Altruistic-Donor Chain." *The New England Journal of Medicine* 360, no. 11 (2009), 1096.

Richerson, Peter J. and Robert Boyd. *Not by Genes Alone: How Culture Transformed Human Evolution* (Chicago: University of Chicago Press, 2004).

Ridley, Matt. *Die Biologie der Tugend. Warum es sich lohnt, gut zu sein* (Berlin: Ullstein, 1997).

Rilling, J. K. et al. "A Neural Basis for Social Cooperation." *Neuron* 35, no. 2 (2002), 395–405.

Rizzolatti, Giacomo and Laila Craighero. "The Mirror-Neuron System." *Annual Review of Neuroscience* 27, no. 1 (2004), 169–192.

Rockenbach, Bettina and Manfred Milinski. "The Efficient Interaction of Indirect Reciprocity and Costly Punishment." *Nature* 444, no. 7120 (2006), 718–723.

Rodriguez-Laso, Angel, Maria Victoria Zunzunegui, and Angel Otero. "The Effect of Social Relationships on Survival in Elderly Residents of a Southern European Community: A Cohort Study." *BMC Geriatrics* 7 (2007), 19.

Roes, F. L. and M. Raymond. "Belief in Moralizing Gods." *Evolution and Human Behavior* 24, no. 2 (2003), 126–135.

Rousseau, Jean-Jacques. *Der Gesellschaftsvertrag oder die Grundsätze des Staatsrechtes* (Frankfurt am Main: Fischer-Taschenbuch-Verlag, 2005).

_____. *Diskurs über die Ungleichheit: kritische Ausgabe des integralen Textes* (Paderborn: Schöningh, 2008).

_____. *Schriften zur Kulturkritik* (Hamburg: Meiner Verlag, 1995).

Rozin, Paul, Jonathan Haidt, and Katrina Fincher. "From Oral to Moral." *Science* 323, no. 5918 (2009), 1179–1180.

Sageman, Marc. *Understanding Terror Networks* (Philadelphia: University of Pennsylvania Press, 2004).

Samuelson, Larry. *Evolutionary Games and Equilibrium Selection* (Cambridge: MIT Press, 1997).

Sanfey, A. G. "The Neural Basis of Economic Decision-Making in the Ultimatum Game." *Science* 300, no. 5626 (2003), 1755–1758.

Scheer, Regina. *AHAWAH, das vergessene Haus: Spurensuche in der Berliner August-straße*, 4th ed. (Berlin: Aufbau-Taschenbuch-Verlag, 2004).

Schrenk, Friedemann. *Die Frühzeit des Menschen. Der Weg zum Homo Sapiens* (Munich: C. H. Beck, 2003).

Schroeder, David. *The Psychology of Helping and Altruism: Problems and Puzzles* (New York: McGraw-Hill, 1995).

Schwarz, James. "Death of an Altruist" in Timothy Ferris, ed., *The Best American Science Writing 2001* (New York: Ecco, 2001).

Segal, Nancy L. and Scott L. Hershberger. "Cooperation and Competition between Twins: Findings from a Prisoner's Dilemma Game." *Evolution and Human Behavior* 20, no. 1 (1999), 29–51.

Seligman, Martin E. P. *Learned Optimism* (New York: Free Press, 1998).

_____. "Why Is There So Much Depression Today? The Waxing of the Individual and the Waning of the Commons." *Contemporary Psychological Approaches to Depression: Theory, Research, and Treatment* (1990), 1–9.

Selten, R. and A. Ockenfels. "An Experimental Solidarity Game." *Journal of Economic Behavior and Organization* 34, no. 4 (1998), 517–539.

Sharp, Michael. *Book of Life: Ascension and the Divine World Order.* (Avatar Publications, 2003).

Sherif, Muzafer. "Experiments in Group Conflict." In Jennifer M. Jenkins et al., eds. *Human Emotions: A Reader* (Oxford: Blackwell, 1998), 245–252.

_____ et al. *Intergroup Conflict and Cooperation: The Robbers Cave Experiment* (Norman, Oklahoma: The University Book Exchange, 1961).

Sherif, M. and C. W. Sherif. *Groups in Harmony and Tension: An Integration of Studies on Intergroup Relations* (New York: Harper and Bros., 1953).

Shermer, Michael. *The Mind of the Market: Compassionate Apes, Competitive Humans, and Other Tales from Evolutionary Economics* (New York: Henry Holt, 2007).

Silk, J. "Who Are More Helpful, Humans or Chimpanzees?" *Science*, no. 5765 (March 3, 2006), 1248.

Silk, Joan B. et al. "Chimpanzees Are Indifferent to the Welfare of Unrelated Group Members." *Nature* 437, no. 7063 (2005), 1357–1359.

Singer, T. "The Neuronal Basis and Ontogeny of Empathy and Mind Reading: Review of Literature and Implications for Future Research." *Neuroscience and Biobehavioral Reviews* 30, no. 6 (2006), 855–863.

_____ and E. Fehr. "The Neuroeconomics of Mind Reading and Empathy." *American Economic Review* 95, no. 2 (2005), 340–345.

Singer T., S. J. Kiebel, J. S. Winston, R. J. Dolan, and C. D. Frith. "Brain Responses to the Acquired Moral Status of Faces." *Neuron* 41, no. 4 (2004), 653–662.

Singer T., B. Seymour, J. O'Doherty, H. Kaube, R. J. Dolan, and C. D. Frith. "Empathy for Pain Involves the Affective but Not Sensory Components of Pain." *Science* 303, no. 5661 (2004), 1157.

Singer, Tania, Ben Seymour, John P. O'Doherty, Klaas E. Stephan, Raymond J. Dolan, and Chris D. Frith. "Empathic Neural Responses Are Modulated by the Perceived Fairness of Others." *Nature* 439, no. 7075 (1, 2006), 466–469.

Singer, Tania et al. "Effects of Oxytocin and Prosocial Behavior on Brain Responses to Direct and Vicariously Experienced Pain." *Emotion* 8, no. 6 (2008), 781–791.

Smith, Adam. *Inquiry into the Nature and Causes of the Wealth of Nations*. Library of Economics and Liberty (http://www.econlib.org/cgi-bin/searchbooks.pl?searchtype=BookSearchPara&pgct=1&sortby=R&searchfield=F&id=10&query=pins&x=0&y=0&andor=and).

_____. *The Theory of Moral Sentiments* (Philadelphia: A. Finley, 1817).

Smith, Vernon L. *Rationality in Economics: Constructivist and Ecological Forms* (Cambridge: Cambridge University Press, 2007).

Sosis, R. "Religion and Intragroup Cooperation: Preliminary Results of a Comparative Analysis of Utopian Communities." *Cross-Cultural Research* 34, no. 1 (2000), 70.

_____ and E. R. Bressler. "Cooperation and Commune Longevity: A Test of the Costly Signaling Theory of Religion." *Cross-Cultural Research* 37, no. 2 (2003), 211.

Spencer, Herbert. *Principles of Biology* (London: William and Norgate, 1864).

_____. *Social Statistics* (London: John Chapman, 1851).

Stallman, Richard. "Initial Announcement." http://www.gnu.org/gnu/initial-announcement.html.

Steinberg, Avraham, MD. *Encyclopedia of Jewish Medical Ethics* (Nanuet, NY: Feldheim, 2003).

Stevens, Jeffrey R. and Marc D. Hauser. "Why Be Nice? Psychological Constraints on the Evolution of Cooperation." *Trends in Cognitive Sciences* 8, no. 2 (February 2004), 60–65.

Stiner, Mary C., Ran Barkai, and Avi Gopher. "Cooperative Hunting and Meat Sharing 400–200 kya at Qesem Cave, Israel." *Proceedings of the National Academy of Sciences* 106, no. 32 (2009), 13207–13212.

Tabibnia, Golnaz and Matthew D. Lieberman. "Fairness and Cooperation Are Rewarding: Evidence from Social Cognitive Neuroscience." *Annals of the New York Academy of Sciences* no. 1118 (2007), 90–101.

Takahashi, H. et al. "When Your Gain Is My Pain and Your Pain Is My Gain: Neural Correlates of Envy and Schadenfreude." *Science* 323, no. 5916 (2, 2009), 937–939.

Takahashi, N. "The Emergence of Generalized Exchange 1." *American Journal of Sociology* 105, no. 4 (2000), 1105–1134.

Tan, J. H. W. and F. Bolle. "Team Competition and the Public Goods Game." *Economics Letters* 96, no. 1 (2007), 133–139.

Tankersley, Dharol, C. Jill Stowe, and Scott A. Huettel. "Altruism Is Associated with an Increased Neural Response to Agency." *Nature Neuroscience* 10, no. 2 (2007), 150–151.

Tapscott, Don and Anthony D. Williams. *Wikinomics: How Mass Collaboration Changes Everything* (New York: Portfolio/Penguin, 2006).

Titmuss, Richard M. *The Gift Relationship: From Human Blood to Social Policy*, expanded and updated edition (New York: The New Press, 1997).

Tomasello, M. "Chimpanzees Understand Psychological States—The Question Is, Which Ones and to What Extent?" *Trends in Cognitive Sciences* 7, no. 4 (2003), 153–156.

Tomasello, Michael et al. *Why We Cooperate* (Cambridge: MIT Press, 2009).

Trivers, R. L. "The Evolution of Reciprocal Altruism." *The Quarterly Review of Biology* 46, no. 1 (1971).

Trivers, Robert. "As They Would Do to You—A Review of *Unto Others: The Evolution and Psychology of Unselfish Behavior*." *Skeptic Magazine* (1998) (http://www.skeptic.com/magazine/archives/vo106n04.html).

Trotter, R. J. "Muzafer Sherif: A Life of Conflict and Goals." *Psychology Today* (September, 1985), 55–59.

Turnbull, Colin. *The Mountain People* (New York: Simon and Schuster, 1972).

Tyler, T. R. and E. A. Lind. *Procedural Justice* (New York: Springer, 2005).

Ueno, A. and T. Matsuzawa. "Food Transfer between Chimpanzee Mothers and Their Infants." *Primates* 45, no. 4 (2004), 231–239.

Uvnäs-Moberg, K. and M. Eriksson. "Breastfeeding: Physiological, Endocrine and Behavioural Adaptations Caused by Oxytocin and Local Neurogenic Activity in the Nipple and Mammary Gland." *Acta Paediatrica (Oslo, Norway: 1992)* 85, no. 5 (1996), 525–530.

Vega-Redondo, Fernando. *Evolution, Games, and Economic Behaviour* (Oxford: Oxford University Press, 1996).

Vrba, Elisabeth. "The Pulse that Produced Us." *Natural History* 5 (1993), 47–51.

Waal, Frans de. *Der Affe in uns: Warum wir sind, wie wir sind*, 3rd ed. (Munich: Hanser, 2006).

_____. *The Age of Empathy: Nature's Lessons for a Kinder Society* (New York: Harmony Books, 2009).

Wallace, Björn et al. "Heritability of Ultimatum Game Responder Behavior." *Proceedings of the National Academy of Sciences of the United States of America* 104, no. 40 (2007), 15631–15634.

Walum, Hasse et al. "Genetic Variation in the Vasopressin Receptor 1a Gene (AVPRIA) Associates with Pair-Bonding Behavior in Humans." *Proceedings of the National Academy of Sciences* 105, no. 37 (2008), 14253–14156.

Warneken, F., F. Chen, and M. Tomasello. "Cooperative Activities in Young Children and Chimpanzees." *Child Development* 77, no. 3 (2006), 640–663.

Warneken, F. and F. Tomasello. "Altruistic Helping in Human Infants and Young Chimpanzees." *Science* 311, no. 5765 (2006), 1301.

_____. "Helping and Cooperation at 14 Months of Age." *Infancy* 11, no. 3 (2007), 271–294.

Wedekind, C. "Cooperation through Image Scoring in Humans." *Science* 288, no. 5467 (2000), 850–852.

Wedekind, Claus and Manfred Milinski. "Human Cooperation in the Simultaneous and the Alternating Prisoner's Dilemma: Pavlov versus Generous Tit-for-Tat." *Proceedings of the National Academy of Sciences of the United States of America* 93, no. 7 (1996), 2686.

Welberg, L. "Mirror Neurons Singing in the Brain." *Nature Reviews Neuroscience* 9, no. 3 (2008), 163.

Welty, Joel. *The Life of Birds* (Philadelphia: W. B. Saunders, 1962).

West, S. and A. Griffin. "Social Semantics: Altruism, Cooperation, Mutualism, Strong Reciprocity and Group Selection." *Journal of Evolutionary Biology* 20, no. 2 (2007), 415–432.

Wilkinson, Gerald S. "Reciprocal Food Sharing in the Vampire Bat." *Nature* 308 (March 8, 1984), 181–184.

Wilson, David and Edward O. Wilson. "Evolution 'for the Good of the Group"—It's Time for a More Discriminating Assessment of Group Selection." *American Scientist* 96, no. 5 (2008), 380.

Wilson, David Sloan. *Darwin's Cathedral: Evolution, Religion, and the Nature of Society* (Chicago: University of Chicago Press, 2002).

_____ and Elliott Sober. *Unto Others* (Cambridge: Harvard University Press, 1998).

Wilson, E. O. *Sociobiology: The New Synthesis* (Cambridge: Belknap Press of Harvard University Press, 1975).

Wright, Robert. *The Moral Animal: The New Science of Evolutionary Psychology* (New York: Pantheon Books, 1994).

Wurz, Sarah. "Variability in the Middle Stone Age Lithic Sequence, 115,000–60,000 Years Ago at Klasies River, South Africa." *Journal of Archaeological Science* 29, no. 9 (2002), 1001–1015.

Young, L. J. et al. "Increased Affiliative Response to Vasopressin in Mice Expressing the V1a Receptor from a Monogamous Vole." *Nature* 400, no. 6746 (1999), 766–768.

Zak, P. J. and S. Knack. "Trust and Growth." *Economic Journal* (2001), 295–321.

Zhao, Z. et al. "Worldwide DNA Sequence Variation in a 10-Kilobase Noncoding Region on Human Chromosome 22." *Proceedings of the National Academy of Sciences of the United States of America* 97, no. 21 (2000), 11354–11358.

Zorrilla, E. P. et al. "The Relationship of Depression and Stressors to Immunological Assays: A Meta-Analytic Review." *Brain, Behavior, and Immunity* 15, no. 3 (2001), 199–226.

Index

Page numbers in *italics* refer to illustrations. Page numbers followed by "n" refer to notes.

A
Ache (people), 122
Adams, Scott, 165
African-Americans, 182–83
Alberti, Leon Battista, 214n10
Allman, John, 89–90
altruism
    as concept, 7–8, 9–10
    defined, 7, 9
    justice and, 117–18
    life expectancy and, 89–90
    morality versus, 158–59
    provisional, 127, 131, 132
    reciprocal, 33–34, 38–39, 100–103,
        185
    sympathy and, 67
    See also specific topics
amygdala, 182
Andes Pact, 158
animals. *See specific animals*
appearance, similarity in, 79
Aris, Michael, 120
Au, 121–22
Australopithecus, 109–10
automobile industry, 204–5
autonomous individuals, myth of,
    209
Autrey, Wesley, 3–5, 8–9, 12
Axelrod, Robert, 27–29
"axial age," 178–80

B
Babylonian Exile, 181, 186–87
bakery customer scenario, 27
banks, 38, 196–97
Basel-Land, Switzerland, 132–33
basketball leagues, 139

Batson, David, 64–65
Battle of the Sexes game, 215n31
Bavarian drinking customs, 150
benefits, 9, 10–12
Berlusconi, Silvio, 223n11
bicycle couriers, Swiss, 124, 218n8
birds, 59, 79, 91
Bischof-Köhler, Doris, 68
blood donors, 134
bluestreak cleanser wrasse, 5
Boesch, Christophe, 108
borderline personality disorder (BPD),
    50–51
Bowles, Samuel, 146, 205, 219n15
brain size, human, 111–12, 114–15
Buber, Martin, 169, 184
Buchan, Nancy, 202–4
Buddha, 176, 177, 184
Buddhism, 177–78, 182, 185–86
Bush, George W., 44–45

C
categorical imperative, 186
Ceaușescu, Nicolae, 63
cheating, antennae for, 30, *31*, 32
children
    altruism toward one's own, 75
    chores, paying for, 135
    communal rearing, 112–13
    fairness, inclination to, 114
    helpfulness, 68–69
    language preferences, 163–64
    maturity, time to, 114–15
    nature of, 87–88
    self-knowledge, 68
    sharing, difficulty of, 97
    social intelligence, 107, 113

chimpanzees
  fairness, lack of interest in, 114
  helpfulness, 69–70
  injured, 107
  intelligence, 106–7
  leopards, sticking together against,
    107–8
  orphaned, 107, 108
  sharing, lack of, 105–6
Chinese ethics, 177–78, 179, 180–81
chores, paying children for, 135
Cicero, 177, 187
circumcision, female, 161
climate swings, 110, 111
cod stocks, 197–98
collectivism, 200
communes, 162
compassion. *See* sympathy
competition, 138–39, 155–57, 166–68
Confucius, 174–75, 177, 179, 180–81,
    182, 184
cooperation, 42–44, 71–72
corporations, transnational, 193–94
corruption, 157–59
Cosmides, Leda, 30, *31*, 32
costs, 9, 10–12
court proceedings, 118–19
crimes and criminals, 54–55, 164–66
cuckoos, 79, 91

D
Darwin, Charles, 3, 12, 14, 17–19, 138,
    212n19
Dawkins, Richard, 15–16
daxing, 149
DeBruine, Lisa, 79
dependence, mutual, 209
de Quervain, Dominique, 128–29
Descent of Man, The (Darwin), 18
Desmond, Adrian, 212n19
Deutschkron, Inge, 169, 170–71, 172
de Waal, Frans, 69
dietary taboos, 161–62
Dinka, 220n21
disgust, 129
diversity, social, 200–201
division of labor, 195
Doctorow, E. L., 218n4

dogs as pets, 73–75, 91, 148
dorsolateral prefrontal cortex, 129–30
drinking customs, Bavarian, 150
Dunn, Elizabeth, 207

E
eBay, 39, 40, 188
economic revolution, 204–6
economics majors, 124
egocentrism
  as concept, 7–8
  defined, 7, 9
  See also specific topics
electroshocks, 60–61, 84
Elisha (Biblical prophet), 181
Emilia-Romagna (Italy), 201, 223n11
emotional susceptibility, 59–62, 66
emotions, regulating, 63
empathy
  cooperation and, 71–72
  defined, 57
  emotional susceptibility versus,
    61–62
  intellectual exchange of roles and, 67
  nature of, 57–58
  rationale for, 62
  stages, 63
  success and, 70–71
  sympathy versus, 66
  turning off/not noticing, 64–65
envy, 160
evolutionary biology, costs and benefits
    in, 11–12
executions, 63

F
fairness. *See* justice
family ties, 14–15, 75–77, 101, 138, 143,
    215n4
famine, 35–37, 39
fear, 84–85
Fershtman, Chaim, 164
financial crisis (2008), 38, 196–97
financial markets, international, 196–97, 198
Folk theorem, 213n20
football fans, 58
Fowler, James H., 58, 146–47
Frank, Robert H., 124

Free Rider Game, 125, *126*, 127–28, 131, 133, 199–200
free riders, 124–25, *126*, 127–28
free software developers, 191–92, 223n4 (chap. 11)
Frey, Bruno S., 132–33
friends, 157–58
front cortex of the cerebrum, 129

G

Gallese, Vittorio, 54–55, 56, 59, 63
game theory, 23–24, 25, 32–33, 213n20
  See also specific games
Gates, Bill, 191, 192
gender differences, 63–64, 89–90
generosity, 85–86, 90, 145–46, 207, 210
genetic ties, 14–15, 75–77, 101, 138, 143, 215n4
Germany, 150, 169–73, 221n2
gestures, mirroring, 58, 214n13
Ghiselin, Michael T., 10
Gibran, Khalil, 20
Gigerenzer, Gerd, 30, *31*, 32
globalization, 203–4
global village, 202–4
Gnau, 121–22
Gneezy, Uri, 135
GNU software system, 191–92, 193
goals, common, 166–68
God, 175
Goddard, Roger D., 51–52
Golden Rule
  as change in perspective, 183–85
  history, 176–81
  limits, 185–86
  reciprocal altruism versus, 185
Good Samaritan parable, 64–65, 184–85
Google, 192, 205
gratification, deferred, 102
greed, 13, 128–30, 208–9
Greek city-states, 152–53, 180, 187
Greek philosophers, 175–76, 182
Greenland Vikings, 161, 162
grocery shopping conflicts, 20–21, 23, 25, 26, 27
group selection, 143, *144*, 145, 188, 219n12, 219nn14–15
guitar, learning to play, 59

Gumz, Frau, 172–73
Gürek, Özgür, 131

H

hackers, 190–91
Haldane, J. B. S., 15, 16, 75–76
Hamilton, William D., 77, 101, 138, 139–40, 141, 143
happiness, 42–44, 85–86, 207, 210
Hardin, Garrett, 99–100
helpfulness, 68–70, 174
Henrich, Joseph, 121–22, 179–80
Herrmann, Benedikt, 199–200
Hillel, Rabbi, 177
Hinduism, 176, 177–78
hitchhikers, 63
Hitler, Adolf, 13, 155
Hitler salute, 58, 63
Hobbes, Thomas, 87, 88
Hoffmann, Roald, 70–71
Holocaust, 169–74, 221n2
Holzach, Michael, 74
Homo economicus, 37–38, 43, 44, 50, 116–17, 119
honesty, taxpayer, 132–33
household conflicts, 20–21, 23, 25, 26, 27
Hrdy, Sarah Blaffer, 112–13
Huang, Chen-Ying, 72
Huettel, Scott, 67
humankind, view of, 87–88, 133–34
humans, early, 95, 96–97, 98–99, 110–11, 112, 113–14
hunter-gatherer societies, 35–37, 39, 159–60
hunting behavior, 95–97, 98–99, 104, 113–14, 122–23, 193
Hussein, Saddam, 63

I

IBM, 192
Ik, 35–37, 39
imitation, learning by, 58–59
income inequality, 34
India, 176, 180, 184
individualism, 200–201
Indonesian whale hunters, 121–23
Industrial Revolution, 196
infants. *See* children

information, trade in, 205–6
initiation rituals, 161–62
insects, 5, 15, 76–77, 78
Insel, Thomas, 81–82
insular cortex, 51
intellectual exchange of roles, 67
interconnectedness, 202–4
Internet, 6–7, 192–93
Islam, 184, 189
Isocrates, 184
Italy, 201, 223n11

J
Jainism, 176
Japanese kamikaze pilots, 166
Jarrah, Ziad, 164–65
Jaspers, Karl, 178, 180
Jesus, 178, 184–85
Jews
    Babylonian Exile, 181, 186–87
    Holocaust, 169–74, 221n2
Judaism, 177–78, 182
justice
    altruism and, 117–18
    charity versus, 117
    children and, 114
    chimpanzees and, 114
    as principle, 118–20, 123–24
    sacrifices for, 118–20
    sharing and, 120
    universal, 187
    See also Free Rider Game; Ultimatum
        (game)

K
kamikaze pilots, 166
Kant, Immanuel, 186
Keefer, Philip, 52
King-Casas, Brooks, 46–47, 50–51
kinship, 14–15, 75–77, 101, 138, 143,
    215n4
Klasies River caves, 95–96, 104, 114
Kleist, Heinrich von, 118, 218n4
Knack, Stephen, 52
Knoch, Daria, 130
knowledge, trade in, 205–6
Kosfeld, Michael, 49
Kropotkin, Prince, 14

L
labor, division of, 195
Lakhani, Karim R., 223n4 (chap. 11)
Lamalera, Indonesia 121–23
Lang, Frieder R., 76
language diversity, 154, 162–64
Lao-Tse, 35
lawyers, corporate, 119
learning by imitation, 58–59
Lehman Brothers, 196
Leonardo da Vinci, 60
leopards, 107–8, 109
Lessing, Gotthold Ephraim, 65, 66
Licht, Alice, 171
life expectancy, 89–90
Lind, E. Allan, 119
Linux, 192
lionesses, 103–5
Liquid Trust, 49–50
Livy, 187
love, chemistry of, 79–81

M
Machiguenga, 121, 122–23
Mafia, 159
Mahavira, 176
Marx, Groucho, 15
massacres, Turkish, 154
McCabe, Kevin, 45, 46, 48–49, 50
medial prefrontal cortex, 68, 71
mice, 81–82
Michael Kohlhaas (Kleist), 118,
    218n4
Microsoft, 191, 192
Milo, Richard, 96
mirror neurons, 55–57, 58, 59,
    60
mirror test, 68
Mokyr, Joel, 196
monkeys, 55–56, 59, 60–61
Moore, James, 212n19
Morales, Oscar, 202
morality
    altruism versus, 158–59
    history, 151–53
    interconnectedness and, 201
    rewards and, 135–36, 218n24
    trust and, 133

universality, 174–76
See also specific topics
mutilation, voluntary, 161–62

N
Nash, John, 25
Nash equilibrium, 25–27
Nazi persecution, efforts against, 169–74, 221n2
Nebuchadnezzar II, 181, 187
Nesse, Randolph M., 33–34
networking, 158, 202–4
Neyer, Franz J., 76
norms, 148–51, 179, 189, 199, 220n21
See also Golden Rule
nuclear warfare, 24, 25, 32–33
Nuer, 220n21

O
Oliner, Pearl M., 173–74
Oliner, Samuel P., 173–74
Ōnishi, Takijirō, 166
opioids, 89
opportunists, 127
orange grove harvesting teams, 157
orangutans, 107
organ donors, 6, 134, 211n5
Ottoman Empire, 154–55
oxytocin, 49–50, 80–81, 83–85, 89

P
Packer, Craig, 104–5
painting, 60, 214n10
Papua New Guinea, 158–59
parking rules, German, 150
Parks, Rosa, 119
Parr, Lisa, 60
Pascal, Blaise, 73
Pavlov strategy, 212n9
Peloponnesian War, 152–53
pension experiment, *31*, 32
pets, 73–75, 91, 148, 215nn2–3
Phelps, Elizabeth, 181–82
philosophers, Greek, 175–76, 182
pigeons, 102
Pink Floyd, 54
Poland, 221n2
Posel, Dorrit, 76

posterior superior temporal cortex (pSTC), 66–67
prairie voles, 81–82
Price, George, 137–41, 145, 153
Price equation, 138, 142, 145–46, 153, 219n2
See also group selection
Princeton Theological Seminary, 64–65
prisoner's dilemma
about, 27–28, 29
kinship ties, 77
Pavlov strategy, 212n9
Tit for Tat strategy, 28, 29, 30
trust game compared to, 43
trust in, 40, *41*, 42–43, 44
property, communal, 198
prosperity, 51–53
protohumans, 109–10
provisional altruists, 127, 131, 132
pSTC (posterior superior temporal cortex), 66–67
public transportation free riders, 125, 127
punishment, 127–32, 147–48, 199–200
Putin, Vladimir, 45
Putnam, Robert D., 201

R
racism, 182–83
Ragtime (Doctorow), 218n4
Rakoczy, Hannes, 149
rats, 80–81
Rebel Without a Cause (film), 25–26
reciprocal altruism, 33–34, 38–39, 100–103, 185
refusal, logic of, 25–26
Regio septalis, 49
regulations, 198
rejection, pain of, 88–89
ren (Confucian principle), 174–75, 221n6
reputation, 188
retaliation, 27–28, 29, 30, 199–200
rewards, morality undermined by, 135–36, 218n24
reward system in brain, 43, 48, 50, 86, 89, 128–29
rhesus monkeys, 55–56, 59, 60–61

Rilling, James, 42–43, 44
risk, 39–40, *41*, 42
Rockefeller, John D., 13
Rockenbach, Bettina, 131
roles, intellectual exchange of, 67
Roman Empire, 187
Rousseau, Jean-Jacques, 20, 21, 87, 88, 98

S
salaries, fair, 118
Samson (Biblical judge), 165–66
savanna, 109–10
scarring, decorative, 161–62
science, 206
sea turtles, 75, 80
Second World War, 166, 167–68
Segal, Nancy L., 77
Selfish Gene, The (Dawkins), 15–16
self-knowledge, 67–70, 184
September 11, 2001, attackers, 164–65
Shakespeare, William, 29, 160, 188
shame, 147–48
sharing
    animals and, 100–106, 217n16
    chimpanzees and, 105–6
    difficulty of, 97–99
    justice and, 120
Sherif, Muzafer, 154–57, 166–67
shocks, electrical, 60–61, 84
Simenon, Georges, 54
Singer, Tania, 62, 63–64, 84
Skilling, Jeffrey, 16
slavery, 17, 212n19
small-claims court, 118–19
Smith, Adam, 195–96, 197, 198, 205
Smith, Vernon L., 217n2
Snyder, Mark, 63
social intelligence, 107, 113
social media, 202
social networks, 90
sociobiology, 14–16, 101, 141–42
Socrates, 175–77
software developers, 191–92, 223n4 (chap. 11)
Sosis, Richard, 162
Sparta (Greece), 152–53
Spencer, Herbert, 13

stag hunt parable, 20, 21, 23, 98
Stallman, Richard, 190–92
Standard Oil Company, 13
Stiner, Mary C., 104
subway rescue story, 3–5, 8–9, 12
success, 51–53, 70–71
suicide missions, 165, 166
summer camp rivalry, 155–57, 166–67
Suu Kyi, Aung San, 119–20
Switzerland, 124, 132–33, 218n8
sympathy
    altruism and, 67
    benefits, 65
    for criminals, 54–55
    cynicism about, 56–57
    emotional susceptibility versus, 66
    empathy versus, 66
    gender differences, 63–64

T
taboos, 161–62, 220n16 (chap. 9)
taxpayer honesty, 132–33
Teresa, Mother, 8
Thornhill, Randy, 141
Tierra del Fuego, 18
Tit for Tat (prisoner's dilemma strategy), 28, 29, 30
Titmuss, Richard, 135
Tomasello, Michael, 68–69, 69–70
Torah, 177
Torvalds, Linus, 192
touch, 85
Tower of Babel, 154, 162–66
trade, 122–23
trade ratio, 194
traffic, commuter, 23–24, 25
tragedy in the theater, 65
"Tragedy of the Commons, The" (Hardin), 99–100
transnational corporations, 193–94
Trivers, Robert, 33
trust
    ability to, 50–51
    abuse of, 38–39
    blind, 48–50
    interconnectedness and, 201
    lack of, 35–38, 196–97
    morality and, 133

in prisoner's dilemma, 40, *41*, 42–43, 44

riches/success and, 51–53

risk and, 39–40, *41*, 42

trust game, 45–49, 50, 51, 79, 146–47, 164

tsunami (2005), 202

Turkish massacres, 154

Turnbull, Colin M., 36–37, 39

turtles, sea, 75, 80

twins, 77, 146–47

U

Ultimatum (game), 116–17, 118, 120–22, 150–51, 179–80, 217n2

V

vampire bats, 101–2, 103

vasopressin, 80, 81–82, 83

vasotocin, 80

Vero Labs, 49–50

Vikings, Greenland, 161, 162

volunteer work, remuneration for, 218n24

von Neumann, John, 22–23, 24, 25, 32–33

W

Wall Street (film), 13

war, 109, 152–53, 166, 167–68, 219n15

Warneken, Felix, 68–69, 69–70

warriors, noble, 12–13

wasps, paper, 78

wealth disparity, 34

Wealth of Nations, The (Smith), 195–96

Weidt, Otto, 169–71, 172–73

whale hunters, Indonesian, 121–23

Wikipedia, 192–93, 223n4 (chap. 11)

Wilkinson, Gerald S., 101–2

Wilson, E. O., 219n12

Wolf, Robert, 223n4 (chap. 11)

World War II, 166, 167–68

Wright, Robert, 56–57

Acknowledgments

THIS BOOK HAS A LONG HISTORY. The first outlines go back more than fifteen years. They grew out of my resistance to the reigning theory of the time: the sociobiologists' insistence on strict gene selection. I feel a debt to all the scientists who have broadened my horizons in the intervening years—not just with their work but often in extensive discussions. Among the many are Ernst Fehr, Raghavendra Gadagkar, Vittorio Gallese, Herbert Gintis, Stephen Jay Gould (†), Sarah Hrdy, Manfred Laubichler, Olof Leimar and the Human Uniqueness Group at the Berlin Wissenschaftskolleg, Axel Meyer, David Rand, Bettina Rockenbach, Tania Singer, Elliott Sober, Robert Trivers, Frans de Waal, and Edward O. Wilson.

The criticisms of Stefan Bauer, Frank Jakobs, Katrin Kroll, and Wolfgang Schneider helped enormously to make the text more lucid and readable. Hermann Hülsenberg's diagrams make essential relationships clear at a glance. Heartfelt thanks to all of them. As always, Peter Sillem was an extremely sensitive, engaged, and understanding editor. My agent, Matthias Landwehr, also provided strong support for this project.

Thanks also to everyone who worked on this English language edition. David Dollenmayer translated my text in the most nuanced and impeccable way. And it was a true pleasure to work with Matthew Lore, Nicholas Cizek, Jack Palmer, and all the wonderful team at The Experiment.

It's almost a cliché for an author to thank his family for their patience and understanding. But in this case, I have especially good reason to do so. There is a strange irony in the fact that a book about devotion to others required so much indulgence from my wife and children. Alexandra Rigos is not just my life's partner but also my colleague and this book, like my previous ones, owes much to her. Without her contributions, my work would not be what it is.

About the Author

STEFAN KLEIN, PHD, has studied physics and analytical philosophy and holds a doctorate in biophysics. After several years as an academic researcher, he turned to writing about science for a general audience. From 1996 to 1999 he was an editor at *Der Spiegel*, Germany's leading news magazine, and in 1998 he won the prestigious Georg von Holtzbrinck Prize for Science Journalism. Today Klein is recognized as one of Europe's most influential science writers and journalists. His interviews with the world's leading scientists are a regular feature in Germany's *Zeit* Magazine. His books, which have been translated into more than twenty-five languages, include the #1 international bestseller *The Science of Happiness, The Secret Pulse of Time,* and *Leonardo's Legacy*. A frequent speaker and university guest lecturer, he lives with his family in Berlin. Translator **DAVID DOLLENMAYER** is emeritus Professor of German at Worcester Polytechnic Institute and the author of *The Berlin Novels of Alfred Döblin*. He is the recipient of the 2008 Helen and Kurt Wolff Translator's Prize and often translates for the *New York Review of Books*.